SpaceShipOne

SpaceShipOne

An Illustrated History

Dan Linehan
Foreword by *Sir Arthur C. Clarke*

ZENITH
PRESS

First published in 2008 by Zenith Press, an imprint of MBI Publishing Company, 400 First Avenue North, Suite 300, Minneapolis, MN 55401 USA.

Zenith Press titles are also available at discounts in bulk quantity for industrial or sales-promotional use. For details write to Special Sales Manager at MBI Publishing Company, 400 First Avenue North, Suite 300, Minneapolis, MN 55401 USA.

To find out more about our books, join us online at www.zenithpress.com.

Front cover: (main) Jim Koepnick/Experimental Aircraft Association; (inset) Mojave Aerospace Ventures LLC, video capture provided courtesy of Discovery Channel and Vulcan Productions, Inc.

Back cover: Courtesy of Mojave Aerospace Ventures LLC, photo by David M. Moore

Page 2: X PRIZE Foundation

Page 6: Courtesy of Virgin Galactic

Library of Congress Cataloging-in-Publication Data

Linehan, Dan, 1969-
  SpaceShipOne : an illustrated history / by Dan Linehan.
     p. cm.
  ISBN 978-0-7603-3188-0 (hardbound w/ jacket)
1. SpaceShipOne (Spacecraft)—Pictorial works.
2. Aeronautics—United States—Records. 3. Space tourism.
4. Space flights. 5. Rutan, Burt. I. Title. II. Title: Space Ship One.
TL795.L56 2008
629.45—dc22
                                        2007043098

Editor: Steve Gansen
Designer: Mandy Iverson

Printed in China

A style note on numbers, units, and conversions: In general, numbers representing data will be provided in English units first and as equivalent values in metric units following within parentheses. However, sometimes important numbers in metric units are given first, as with the Ansari X Prize altitude goal. The exact conversion factors used were obtained from the National Institute of Standards and Technology (NIST). Careful attention to rounding and significant figures of both the original data and the converted data had to be considered. In many cases a balance was struck between scientifically accurate and precise numbers with proper use of significant figures and numbers that provided a more meaningful physical representation. All this is best illustrated in the following explanation of Mach number.

Mach number is the ratio of velocity to the speed of sound. The speed of sound is not a constant, though. It depends on pressure, which in turn is dependant on altitude. An aircraft flying at Mach 1 is flying exactly at the speed of sound. At sea level, the speed of sound is 761 miles per hour (1,220 kilometers per hour), but at 47,000 feet (14,330 meters), the speed of sound is only 660 miles per hour (1,060 kilometers per hour). So, Mach number cannot be converted to velocity unless an altitude is given. Mach number and velocity are only represented together in this book when both values were available from the source of the data.

"I don't believe in taking foolish chances,

but nothing can be accomplished without taking any chance at all."

—Charles A. Lindbergh, *The Spirit of St. Louis*

"If I have seen further, it is by standing on the shoulders of giants."

—Sir Isaac Newton, letter to Robert Hooke

"We have lingered long enough on the shores of the cosmic ocean.

We are ready at last to set sail for the stars."

—Carl Sagan, *Cosmos*

Contents

Acknowledgments

I want to apologize first off to any of my former English teachers who will have heart failure upon hearing that I actually wrote a book. I think I spelled my doom when I put a poem of mine on the first page of my master's thesis. It wouldn't have been so bad if my thesis was about art or literature, but it was about making superconductors with lasers and analyzing them with x-rays and electron beams. Fortunately, this book I write now is not a literary masterpiece, so, any such resulting cardiac irregularities shouldn't last long or will perhaps be particularly mild.

With that disclaimer out of the way, I'd like to get started thanking some pretty fantastic people because without them I would have never been able to pull off a book like this. It has been a long and not so straightforward journey to get to this point. There are more people to thank than I can reasonably do so on this page. It's funny how things in life sometimes fall into place after a chain of events occurs that no one could have ever predicted in a million years. That's what it's all about, I guess.

I started writing for the California International Airshow's event program in 2003 thanks to Cindy Rogers. I soon became editor and publisher working for Ginny Brown, who stood behind me and allowed me to create an award-winning event program in just three years. But I was far from alone in this. I met Tyson Rininger during this time. His wonderful photos filled the event programs from cover to cover. It was Tyson who got me on board with MBI. It is impossible to express my full appreciation. I am very fortunate to have been able to include Tyson's photos of *SpaceShipOne*'s first spaceflight and trip to Oshkosh.

I had the opportunity to write about *SpaceShipOne* and the Ansari X Prize for the Airshow. Lilian, Ray, and Cheryl hooked me up with VIP access during X2, which gave me the chance to met Burt Rutan, Paul Allen, Peter Diamandis, Anousheh Ansari, and Sir Richard Branson. In addition, this book would be nowhere if it wasn't for the tremendous support of Dave Moore from Vulcan. Dave was Paul Allen's managing director for Tier One. He was there from the beginning for the construction and test flights of *SpaceShipOne* by Scale Composites. Vulcan contributed a majority of the images used in this book, and Dave shot many of them.

When I made the transition from engineering to writing, David Gitin, a creative writing instructor at Monterey Peninsula College and an incredible poet, was critical to my successful transition. His friendship has been invaluable. I left engineering in 2000 with one goal in mind, to write a book. Along the way, not one of my family and friends told me I was mad. They may have thought it, but at least they kept it to themselves. Mom, Dad, Steve, Rob, Jen, Joe, James, Vik, Patty, Zaheer, Pranita, Norma, Charo, Maria, Manuel, Kim, Jerrold Kortney, and, of course, the Princess of the Bottom of the World, thank you so much for understanding.

Last, it's very difficult to express my amount of gratitude toward MBI and my editor Steve Gansen. Writing this book was more challenging than I could have ever envisioned. Steve was vital in guiding me through the process and helping me reach a dream.

My best thoughts and wishes to you all.

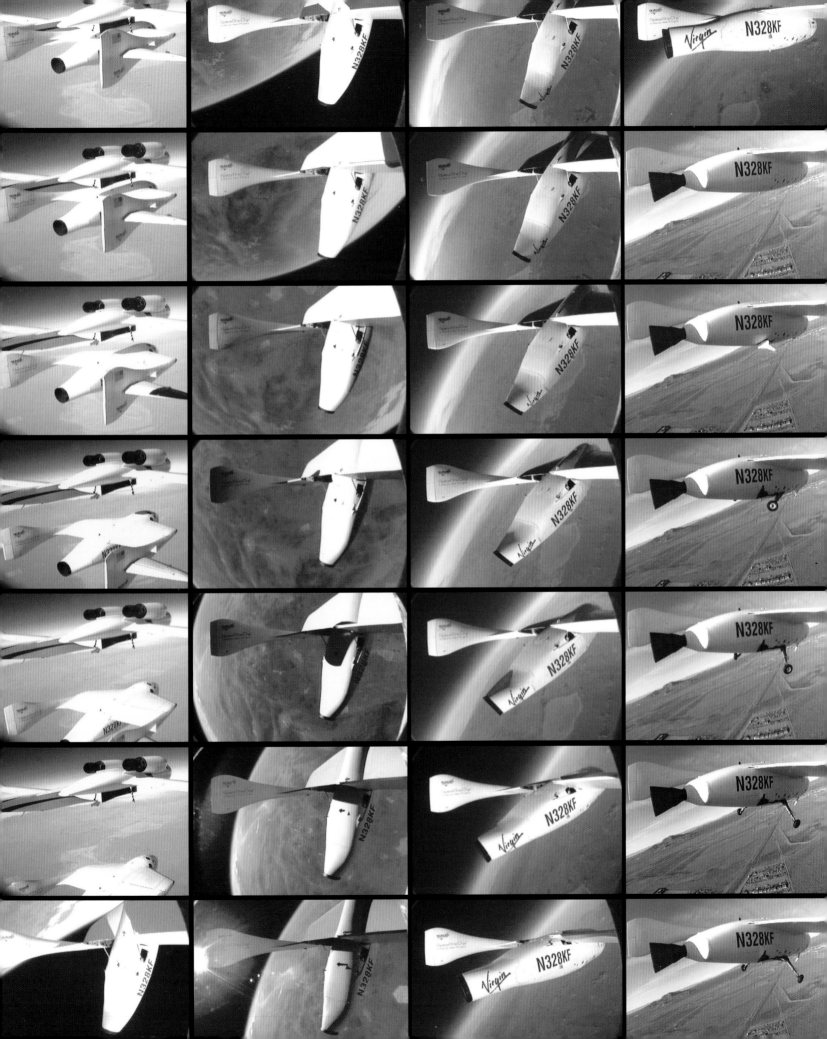

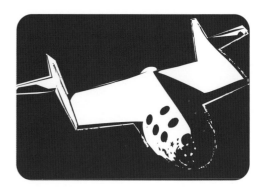

Foreword

Enter Citizen Astronauts

Escaping from Earth will not always be astronomically expensive; contrary to the impression created by a Saturn launch, the energy needed to reach space is remarkably small.

"About eight hundred pounds of kerosene and liquid oxygen, costing some twenty-five dollars, will liberate enough energy to carry a man to the Moon. The fact that we currently burn a thousand tons per passenger indicates that there is vast room for improvement.

"This will come through the space refueling, nuclear propulsion and, most important of all, the development of reusable boosters, or 'space ferries,' which can be flown for hundreds of missions, like normal aircraft. We have to get away, as quickly as possible, from today's missile-orientated philosophy of rocket launchers which are discarded after a single flight."

When I wrote these words in July 1969, the Apollo 11 astronauts were on their way to the Moon.* The Space Age was barely a dozen years old, and travelling to space required so much money and effort that only the governments of the richest countries could engage in it.

But the early years of space exploration were driven by different considerations: both national space agencies and TV networks seemed to love the massive firework displays at rocket launches. Indeed, witnessing a Saturn V take off could be a moving experience—if we overlooked the fact that it was a grandiosely wasteful way to travel anywhere (each Apollo mission cost around two hundred million in 1960s dollars).

So, even as I covered the Moon Landings for CBS television, I was already looking ahead to a time when space travel would make more economic sense. In my essay, I envisaged that the true Space Age would dawn sometime after 1985, ". . . and projects which today are barely feasible will become not only relatively easy, but *economically self-supporting*."

I added: "The closing years of this century should see the beginnings of commercial space flight, which will be directed first towards giant manned satellites or space platforms orbiting within a thousand miles above the Earth's surface."

Well, in those heady days of Apollo, I couldn't have anticipated all the detours and distractions of the 1970s that delayed our optimistic projections. Politics and economics have taken their toll, but looking back, I'm happy to note that I was off by only a decade or so.

Commercial space flight is now beginning to be technologically feasible and will soon become economically viable. The rise of citizen astronauts has already begun—this time, I doubt if politics can hold up progress because it is no longer so closely tied to the fluctuating interests and resources of national governments.

*See Chapter 10 in my collection of essays, *The View from Serendip* (Random House, 1978).

SpaceShipOne: An Illustrated History chronicles a key milestone in the race to take private citizens and private enterprise to space. It's the story of how a group of determined and passionate aerospace designers—and their financiers—pulled off one of the most remarkable accomplishments in our conquest of gravity.

In that process, they won the ten-million-dollar Ansari X Prize founded by my friend Dr. Peter Diamandis to galvanize private enterprise and technological innovation in space travel. The prize was modeled after the twenty-five-thousand-dollar Orteig Prize, offered in 1919 by hotelier Raymond Orteig, to the first pilot to fly nonstop between New York and Paris. An unknown airmail pilot named Charles Lindbergh finally won this challenge in May 1927, flying a single-engine aircraft named *Spirit of St. Louis*. That feat won him instant fame and spawned the commercial aviation industry that changed our world beyond recognition.

And here's an interesting coincidence. In 1987, I received the Lindbergh Award presented annually for those seeking a balance between technology and nature. The winner in 2000 was Burt Rutan, who went on to design *SpaceShipOne* with generous backing from Microsoft co-founder (and science fiction enthusiast) Paul Allen.

Burt and I are connected in other ways. I am intrigued to read that Burt's boyhood imagination was sparked by watching Wernher von Braun on TV talking about the exploration of the Moon and Mars. Wernher was a good friend who took to diving on my suggestion—I told him that the best simulation of weightlessness was achieved underwater. And a few years ago I found out, from his long-time secretary, how Wernher had used my 1952 book, *The Exploration of Space*, to convince President Kennedy that it was possible to land men on the Moon.

While writing this, I came across Burt's remark in *Popular Mechanics* (September 2007): "If we make a courageous decision like the goal and program we kicked off for Apollo in 1961, we will see our children or grandchildren in outposts on other planets."

Fortunately, we need not rely solely on governments for expanding humanity's presence beyond the Earth. The Ansari X Prize has succeeded in spurring commercial astronautics, and I hope governments will not stand in the way. I am following with much interest the emergence of a new breed of "astropreneurs" who are trying out new technologies, business models—and indeed, building a whole industry—without relying on government funding.

In that sense, space travel is returning to where it started: with maverick pioneers dreaming of journeys to orbit and beyond, some carrying out rocket experiments in their own backyards. Burt and his team have been a great deal more successful than Robert Goddard ever was in his lifetime (and, thankfully, no one is ridiculing Burt the way they did with Goddard).

Yet, today's astropreneurs like Paul Allen and Burt Rutan are driven by the same spirit of enquiry, adventure, and exploration that sustained Lindbergh and Goddard. This, then, is the inside story of how citizens reclaimed space.

—Sir Arthur C. Clarke
Colombo, Sri Lanka
4 October 2007

Doug Shane (left) and Burt Rutan (right) give a peek inside the secret Tier One space program at *SpaceShipOne* and *White Knight* during the very early stages of construction. *Mojave Aerospace Ventures LLC, photograph by Scaled Composites*

Chapter 1

A Secret Space Program

When *SpaceShipOne* touched down on Mojave's Runway 30 after its third flight into space, it not only won the Ansari X Prize, but it also, in a way, completed a journey that had started on two separate paths that intertwined along the way and then finally merged together in 2004. *SpaceShipOne* was a small, lightweight rocketship somewhat resembling NASA's early X-planes but with elements of Burt Rutan's distinct flair for the unique and unconventional. Figure 1.1 shows Burt Rutan with Doug Shane, the test flight director, and the three *SpaceShipOne* test pilots, Pete Siebold, Brian Binnie, and Mike Melvill.

On that day test pilot Brian Binnie did more than capture the Ansari X Prize. He captured people's imagination and reignited the space-crazy in them. Back in 1927, Charles Lindbergh, who just barely cleared telephone lines after takeoff in pursuit of the Orteig Prize, crossed the Atlantic Ocean nonstop from New York to Paris and sparked a boom in aviation like none other. These two turning points forever changed the way people looked up into the sky and saw themselves flying free as the birds or high as the stars.

Without people there would be no flying machines. It is not the mechanisms of engine, fuselage, wing, and empennage that provide transport into the air and through the clouds: it is the people whose ideas, visions, and daydreams have taken flight and soared. Without the human mind, the greatest height reached would only be as high as the highest surface that could be climbed.

Fig. 1.1. Burt Rutan (top left), the *SpaceShipOne* and *White Knight* designer, and Doug Shane (top right), the test flight director, stand behind their test pilots, Pete Siebold, Brian Binnie, and Mike Melvill (left to right). *Mojave Aerospace Ventures LLC, photograph by Scaled Composites*

It is no coincidence that the breakthroughs in aviation have come from ingeniousness and inventiveness as early as Leonardo da Vinci's drawings of flying corkscrews and birdlike flight suits, or even earlier when the ancient Greeks pondered the air around them. As ideas like these were being conceptualized, more often than not, peers would view such thinking as folly, insanity, or even sacrilege.

Thankfully, there are a few whose hides are thicker than most, whose resilience is more enduring than most, and whose passion burns brighter than most. And more importantly, there are some whose persistence in moving forward, even if it takes stepping backward at times, is unwavering. It is not enough to be a great thinker. The genius lies in the execution. As inventor Thomas Edison is famous for saying, "Genius is one percent inspiration, ninety-nine percent perspiration."

But these factors are no longer enough in this day and age. In centuries past, it may have been sufficient for just one person to cultivate a dream, from beginning to end, into fruition. But in today's ever changing, ever more complicated society, often a wheel with only one cog will not provide the grip needed to take hold of an idea and spin it into something magical. It is only when the sprocket has gathered enough teeth that it can truly turn and move forward in a synchronicity of movement.

Burt Rutan

In 1965, Elbert L. "Burt" Rutan found himself in the backseat of an F-4 Phantom trying to figure out how to regain control as the fighter jet whirled around in circles in an unrecoverable flat spin. At the time, the Phantom was one of the U.S. Air Force's frontline fighters. Under the right set of conditions, though, pilots could enter a flat spin where the only way out required the use of their ejection seats. A total of sixty-one aircraft had been lost because of this, and Rutan's job was to determine a way to recover from this disastrous condition.

Fresh out of college, it was certainly a dream job for someone who had craved a challenge. Born in Portland, Oregon, in 1943 and raised in Dinuba, California, southeast of Fresno, he had been flying model airplanes that he designed and built since he was a young boy.

"None of those things were kits," Rutan said. "They were original designs." He entered model-airplane competitions during high school while also learning to fly. His models took him to the nationals, and he brought home trophies.

"I was of course fascinated by space," Rutan recalled. "I listened to the Alan Shepard flight on the radio as I was driving to my college to interview."

Rutan attended California Polytechnic University, where he would receive his bachelor of science in aeronautical engineering. In 1964, he was one a very few students, the only one from his college, selected to attend the Cal Tech Space Technology Summer Institute. During this time the United States continued to lag behind the Soviet Union in the space race. Yuri Gagarin, aboard Vostok 1, had become the first man in space three years earlier, and it would be yet another five years until Neil Armstrong and Buzz Aldrin would set foot on the Moon.

The U.S. space program was hungry for engineers during this electrified time. It seemed natural that someone like Rutan would have his sights set on Apollo manned Moon missions.

Rutan remembers his boyhood imagination being stirred by watching television programs where Wernher von Braun, who headed Germany's V-2 program but was now leading NASA's Saturn V rocket development, talked about the exploration of the Moon and Mars with Walt Disney.

"Von Braun was a big hero of mine because of the Disneyland television show in 1955," Rutan said. "Those programs were enormously compelling to me as a twelve-year-old because we didn't know much about Mars in '55. I have a college astronomy textbook that I had gotten a long time ago. It was written in '53. It is interesting to read because they are debating what kind of life is likely to be on Mars and would there be a chance that it would be intelligent life. So, imagine yourself back in a time period when you believed there was vegetation there because you saw the colors change in the telescopes."

Rutan had a tough decision to make when it came time to leave college. Although the space program barreled forward, and opportunities certainly waited for him, he was very skeptical about how and where he would fit in. "I felt I was so far behind on being able to come in and take a new idea and actually get it out there flying if I focused on spaceflight or manned spaceflight."

He didn't want to work on the space widget of the what-cha-ma-call-it subsystem. While this obscure part was indeed a piece of the puzzle that would have been necessary for launch, Rutan desired to really make a big impact and influence the big picture. Working for an airliner or fighter manufacturer didn't interest him, for the same reason. "I'd be working on a bulkhead or a door."

So, he turned in another direction. "I thought I could make a big difference with general aviation, which I thought was archaic and frozen." He felt that he'd have the opportunity to let his creativity fly, even if it wouldn't be quite as high as if he worked for the U.S. space program. But Rutan's competitiveness helped sway him in yet another direction.

"I couldn't bring myself to go work on Cessnas while everyone I went to school with was on the way to the Moon. So, I made a compromise. I went into air force flight testing."

This was by far the best decision Rutan could have ever made. Working as a civilian at Edwards Air Force Base in Mojave and spending six to ten months on an aircraft before moving to the next one, he learned what risks to take and what decisions to make when testing out new aircraft. This education, he felt, was critical for a designer.

"I'm out there evaluating the performance, flying qualities, safety, etc., of the top-of-the-line, brand-new military airplanes. I got to fly in them, measure data, and report on their performance."

In the Phantom on that day in 1965 when it went into a flat spin, once Rutan and the test pilot were sure that the flat spin was unrecoverable, they deployed a special recovery parachute that forced the large fighter jet out of the spin. "I've done the only flat spin in an F-4 that did not have an ejection." But the very next flight, with a different test engineer in the backseat, the spin recovery chute failed, and both the pilot and test engineer ejected to safety. Rutan still keeps in his office a piece of the F-4's canopy from the wreckage that actually has his name on it. Eventually a procedure would be devised using the Phantom's existing landing drag chute, which greatly reduced the rate of unrecoverable flat spins.

Rutan left the air force in 1972. He explains, "Now I could really exercise with my own responsibility, my own authority, my own decision on risk taking with nobody to answer to in developing the VariViggen, the VariEze, the Defiant, the Solitaire, and all these homebuilts. That path could never have accelerated at that rate if I had gone into the space program."

Homebuilts and the Step toward Space

After working two years as the director of Bede Test Center for Bede Aircraft, makers of the BD-5J Microjet, Rutan was ready to go it out on his own. In 1974 he founded Rutan Aircraft Factory (RAF) to continue the development of his first airplane, which he began designing back in college and started flying in 1972, and the development of brand new designs.

"I name very few of my airplanes," Rutan said. "Even the VariViggen, my first one, was named by a guy at work. I told him I was going to build something that was a lot like the Swedish Viggen, except that it had the feature of variable camber that changed the loading. He said, 'Okay, how about VariViggen.'" Rutan simply replied, "Okay."

Although Burt Rutan is well known for the use of composites, the VariViggen's only use of composites was a fiberglass cowling. The aircraft was made out of wood, and its wings were aluminum. However, it had what would become a hallmark of Rutan's—a canard. The canard was a small wing forward of the main wing that enhanced lift, allowed for slower maneuvering speeds, and helped prevent stalling. The VariViggen was powered by a pusher propeller at the back of the aircraft and had vertical fins at the wingtips called winglets, which improved the rate of climb and cruising speed. These two features also reoccurred in many of Rutan's designs. Figure 1.2 shows some of Rutan's earliest aircraft.

Rutan had begun construction of the VariViggen in his spare time while still working for the air force, and it took him four and a half years to complete it. With only seventy-five hours on the aircraft, Rutan flew the VariViggen to Oshkosh, Wisconsin, to give the public its first viewing during the 1972 AirVenture, the annual fly-in held by the Experimental Aircraft Association (EAA), which is the premiere aircraft organization for homebuilts and experimental aircraft. The crowd thrilled at the futuristic-looking aircraft.

"When you look at the wide variety of his designs from the VariViggen to *SpaceShipOne* and everything in between," said Tom Poberezny, the President of EAA, "Burt has been a design leader. He's always been creative and pushing the envelope.

"If you attend a forum at Oshkosh, it is always a packed house because people want to find out what's the latest and greatest in Burt's fertile mind. And that becomes motivational for people in terms of the excitement of being inside of design theory, design thinking, and innovation."

Rutan gave the VariViggen the tail number N27VV. Characteristic of Rutan, the tail numbers of his aircraft weren't arbitrary. All U.S. aircraft begin with *N*, but the "27" stood for the model number Rutan gave it, and *VV* obviously stood for VariViggen.

Another revolutionary aircraft made its debut at the 1972 AirVenture. The tiny little Rand Robinson KR-1—at a length of 12 feet 6 inches (3.8 meters), a wingspan of 17 feet 2 inches (5.2 meters), and a weight of 310 pounds (140 kilograms)—generated a lot of strange looks and disbelief. Ken Rand's homebuilt aircraft had started people talking and thinking a little differently about the way airplanes were built. The outer wing panels, vertical and horizontal tail surfaces, and other parts of the aircraft were made using polystyrene—the same stuff foam coolers and coffee cups are made.

Even though plywood, which is technically a composite because it is made of wood layers glued together, was also used to build the fuselage, the KR-1 made use of composites like no other aircraft of its time. Polystyrene is very light, but it is also very weak. So, epoxy and cloth made of Dynel, a synthetic fiber, covered the polystyrene in order to give it the rigidity and strength needed to hold together. The KR-1 had a profound influence on the evolution of large-scale use of composite materials in homebuilt aircraft.

Rutan's next design, the VariEze, also a pusher prop with a canard and winglets, took him only three and a half months to complete by taking the use of composites to a whole new level. "I didn't have as a goal to build better, lighter, safer, more affordable composite structures. I saw it as a way that I could take a complex aerodynamic shape and build it quickly and get it into flight test," Rutan said.

Fig. 1.2. The early aircraft designs and projects of the Rutan Aircraft Factory (RAF) came in many shapes and sizes, including *Voyager*, *AMSOIL Racer*, Quickie, Defiant, VariViggen, *Grizzly*, NGT, Long-EZ, AD-1, Catbird, VariEze, Boomerang, and Solitaire. *Provided courtesy of Scaled Composites*

"The KR-1 was a wooden airplane with Dynel and foam to shape its outer shape. Its primary structure was wood—the fuselage and wings and tail—whereas if you took the wood out of it, it would fall apart. The VariEze was different. Its composite was its primary structure."

The composite Rutan had used was polystyrene foam sandwiched between layers of fiberglass. "I started my composite work with moldless methods that I used on the VariEze and the Defiant and a bunch of our airplanes," Rutan said. "I did that by copying what they did when they did repairs to molded European sailplanes. European sailplanes were made, and still are, in molds."

But to fix damage to one of these sailplanes, it wasn't necessary to go back to the mold. "The brainstorm that I had was, 'Wait a minute, I could build a whole airplane with these repair methods.' And that's how I came up with hotwire wing cores and the hand-carved foam for the fuselage box. And sure enough, I built a whole airplane without a mold." Figure 1.3 shows Rutan with a composite panel during the assembly process.

Rutan originally had no intention to sell the VariEze. He built it as a research aircraft to do more testing with the canard concept. But during its introduction to the public at EAA's 1975 AirVenture, the public went wild for the design. Rutan responded by slightly enlarging the design and selling the VariEze as a kit airplane, so homebuilders could purchase plans and components to build it themselves.

"The VariViggen, the VariEze, all these designs, were unique in the fact that they were out of character from the type of design that was typical of the day," Tom Poberezny said. "He breaks the mold every time he does something." Figuratively speaking, of course.

In 1986, Rutan's *Voyager* made the first nonstop flight around the world without refueling. It lifted off from Edwards Air Force Base with its tanks full of 7,011 pounds (3,181 kilograms) of fuel, circumnavigated the globe without once landing, covering a distance of 24,986 miles (42,212 kilometers), and then touched down nine days later at Edwards Air Force Base with 106 pounds (48 kilograms) of fuel to spare.

The amazing strength-to-weight ratio of the graphite fiber and honeycomb composite that Rutan used to build *Voyager* allowed a wingspan of 110 feet 8 inches (33.8 meters) and a primary structure weighing in at only 939 pounds (426 kilograms). Fuel actually accounted for 72 percent of its gross takeoff weight.

It was hard not to notice Rutan's design and engineering prowess. And soon NASA, defense contractors, and large aircraft manufacturers began knocking at the door. In 1982, the year construction actually began on *Voyager*, Rutan founded Scaled Composites to specialize in prototype development, offering design all the way through flight testing of full-scale vehicles or scaled-down versions and models. Figure 1.4 shows early designs that Scaled Composites worked on, including the

Starship, Pegasus, and the *Pond Racer*. Scaled Composites built some of these designs from start to finish, but for others the company only contributed to part of the construction.

"Things got more conventional," explained Rutan. "I mean the way we build airplanes. They are still sandwich. They tend to be more honeycomb core than foam core. They tend to prepreg, and we cure them at higher temperatures. We still do them without the autoclave. And we occasionally come up with a breakthrough in manufacturing methods."

But the manufacturing processes, while important, were never the driving force. As always, it was the design and the need to fly that drove the manufacturing processes. And after about thirty years from the time Rutan graduated college, he started to wonder what was next. Space was never that far from Rutan's mind, or reach. In his bookshelf was a fat binder that contained original writings of Wernher von Braun. It was the Mars project. Rutan decided, "'Damn it! Nobody else is going to do this. I'll do something for space.' That didn't dawn on me until about '93 that if I focused on suborbital, then I could do something interesting."

Early Spaceship Designs

"I always had the space bug, keep in mind," Rutan said. "When did I jump in and do it myself? It came in a time when I thought I could do it. And it wasn't with *SpaceShipOne* at first. I was going to do a capsule, and launch it from an airplane that did a steep climb and a parachute recovery." He began sketching out ideas in 1993.

"I was going to build something to fly out of the atmosphere. I'm not saying the things that I originally laid out were easy. They weren't easy, but they weren't really innovative. They were pretty straightforward. It didn't require anything that was new or patentable or breakthrough in nature."

His first ideas focused on a single-person rocket carried as an external store by a mothership. The carrier aircraft would pull up and then shoot it off like a missile.

"Around 1995 or so, we were designing and starting to build an aircraft called *Proteus*," said Doug Shane, the vice president of business development and the first engineer that Rutan hired at Scaled Composites. An award-winning test pilot himself, Shane also became director of flight operations, which is a position he has held at Scaled Composites since 1989. Figure 1.5 shows Shane in Mission Control with Rutan at his side.

Proteus was a utility aircraft designed to fly at high altitudes above 60,000 feet (18,290 meters), but one of its design requirements was to launch a single-person suborbital rocket. *Proteus* would perform a zoom maneuver at 27,000 feet (8,230 meters), pulling up to 40 degrees to assist the rocket's trajectory. Figure 1.6 shows the original launch concept with the rocket attached by an offset mounting, and figure 1.7 shows its separation from *Proteus*.

When Rutan got word of the formation of the X Prize, well before it was called the Ansari X Prize, he decided to change the design from a single-person capsule to a three-person capsule, which was a condition set forth by the X Prize. Figure 1.8 shows the rocket during ascent and the capsule and booster parachuting down after reentry.

The most important part of the design of the spacecraft was the feather, the little protuberances pointed upward away from the blunt end of the capsule. Acting and looking like the "feathers" of a

Fig. 1.3. Burt Rutan began his pioneering work with composites on the VariEze, the second airplane he designed and built. In 1975, the VariEze took him three and a half months to construct, whereas his first airplane, the VariViggen, took more than four and a half years. *Mojave Aerospace Ventures LLC, photograph by Scaled Composites*

badminton shuttlecock, the design would accomplish two very important things. First, the high drag would decelerate the capsule very quickly, which would tremendously reduce the thermal loading and the dangers of heat buildup during reentry. And second, it would act to self-right and stabilize the capsule, orienting the capsule's blunt side down no matter what the attitude during initial reentry. Rutan would call this his "carefree" reentry. This allowed the spacecraft to reenter at a near-vertical trajectory. The Space Shuttle, in comparison, had to precisely control its attitude during reentry, too shallow and it would skip off the atmosphere, too steep and it would experience catastrophic heating.

The capsule would then float down over the water. "I wanted to pick it out while it was still under the chute with a helicopter," Rutan said.

The method was very similar to the way that film canisters ejected from the very first spy satellites were recovered during Project Corona, the joint effort between the Central Intelligence Agency (CIA) and U.S. Air Force. A film canister would parachute down, and an aircraft with a large catch would swoop by and snag it out of the sky.

"My baseline at first was that if it went into the water it was an unexpected failure or emergency. I wanted to grab them and helicopter them back to the launch site," Rutan said.

Something about a parachute and helicopter blades make it seem like a bit of an odd combination, though, even for Rutan. "When you work with parachute recovery," Rutan mused, "you take certain generic risks that you just can't get around. And I thought I could do it. I think it was a much bigger decision that led me to develop something that could survive a steep reentry and land on a runway and not have to be controlled in attitude during reentry.

"That concept, that design, that approach was much more significant."

So, the ideas of a capsule and a parachute were scrapped. But air launch was still on the table. Scaled Composites would also stick with

Fig. 1.4. Scaled Composites built complete aircraft as well as components for many vehicles. Just a few of these are Pegasus, Starship, Triumph, ATTT, ARES, and *Pond Racer. Provided courtesy of Scaled Composites*

using the name "feather" even though the final design looked more like a pair of broken wings.

"If you can buy some time in the event of a propulsion problem instead of being this high off the ground when the motor quits or does something funny," said Shane while sticking his hand out to his side about four feet off the ground, "if you are at 50,000 feet [15,240 meters], you really have a lot more time to deal with it. It gives you an awful lot more options."

Early models, like those shown in figure 1.9 and figure 1.10, used a high-drag speed brake and large deflecting elevons as the feather mechanism for reentry. And Rutan's team demonstrated several models where the feather worked subsonically. But when it came to supersonic speeds, like those encountered during reentry, they didn't trim supersonically. In other words, they were uncontrollable. If the center of gravity (CG) was moved back, they would fly okay, but then they became unstable at subsonic speeds. A computerized fly-by-wire flight control system might have made them flyable, but the cost was too high and the reliability too low.

This was an enormous setback. Only two manned, winged vehicles had made it to space, the X-15 and the Space Shuttle. Both had fatal accidents during reentry. An X-15 had a flight-control failure in 1967, and the Space Shuttle *Columbia* had a failure of its thermal protection system in 2003. It was necessary to solve this problem before any more forward progress could be made.

"Finally, maybe it was some spicy food he ate late at night or some other kind of epiphany," Shane said about Rutan. "He suddenly realized that maybe the thing to do would be to pivot part of the wing and the tails on a hinge on the body and provide that high-drag configuration. And that ended up satisfying all the subsonic and supersonic aerodynamic challenges. That was actually a pretty clever concept, and that's what we went forward with."

Figure 1.11 shows the new spacecraft design with the feather deployed for "carefree" reentry.

Paul Allen

Up to this point, design and development had been relegated to computer analysis and foam models. Burt Rutan hadn't been ready to approach anyone for funding until he felt ready that he could deliver what he'd promised. Figure 1.12 shows Rutan and aerodynamicist Jim Tighe during early computer analysis.

For the design of *SpaceShipOne* and *White Knight*, Scaled Composites would rely heavily on computer analysis because the vehicles would not go through wind-tunnel testing. Figure 1.13 shows an evaluation of *SpaceShipOne* as the tail booms bend downward.

Around this time, Rutan and philanthropist Paul Allen, who cofounded Microsoft in 1975 with his high school friend Bill Gates, had begun exploring the possibility of using high-altitude airplanes circling over Los Angeles as a way to provide broadband wireless to the city. "My first couple of meetings with Paul were not about space at all," recalled Rutan. "There was an interest that he had in something else I was doing. It was related to *Proteus* for telecommunications." They eventually got around to talking about space, and Rutan's idea for a very low-cost suborbital spacecraft. Allen turned out to be a bit of a space enthusiast and became quite interested. But Rutan was still not comfortable with his design, which was still based upon a capsule and parachutes at the time.

Once fiber optics took off, using an airborne telecommunications platform was no longer feasible, but Rutan hadn't stopped thinking about the spaceship. "I figured out the 'carefree' reentry, and I thought I could have something that could land as a glider, be more operable, and a lot safer. I didn't know that 'carefree' reentry would work. I just had a good feeling about it."

In the spring of 2000, Rutan felt he was ready for funding. "I actually asked for a meeting with Paul. And I said, 'Listen, I think

Fig. 1.5. The first engineer that Burt Rutan hired at Scaled Composites, Doug Shane, was responsible for the flight testing of *SpaceShipOne* and *White Knight*. During the flight tests, Shane's was the cool, calm voice on the Mission Control side of the radio. *Mojave Aerospace Ventures LLC, photograph by David M. Moore*

Original Launch Concept

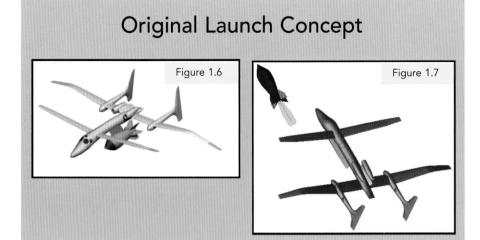

Figure 1.6

Figure 1.7

Fig 1.6. Before the design of *SpaceShipOne* was conceived, Burt Rutan developed a concept for a single-seat rocket to be launched into space off of *Proteus*, a high-altitude research aircraft.

Fig. 1.7. As *Proteus* reached launch altitude, it would perform a zoom maneuver by pointing up at a steep angle to assist the trajectory of the rocket on its suborbital flight. *Mojave Aerospace Ventures LLC, provided courtesy of Scaled Composites*

I could do this now.' And he put out his hand and shook it and said, 'Let's do it.'"

Like many kids growing up in the late 1950s and 1960s, Allen remembered the television cart being wheeled into his classroom to watch Mercury, Gemini, and Apollo launches. Science and technology had fascinated him whether he was building model rockets or reading science fiction. "I always had in the back of my mind, would I ever have the opportunity to do something in a space-related initiative?" Allen recalled. "And so when the *SpaceShipOne* opportunity came up, I was very excited to pursue it."

Paul Allen's company, Vulcan, Inc., and Scaled Composites began a partnership called Mojave Aerospace Ventures. Although Allen and Rutan were aware at the time of the creation of the X Prize by Peter Diamandis, their initial goal, however, was getting to space and not necessarily winning the X Prize.

"None of these meetings were about the X Prize," Rutan said. "People think we did the program for the X Prize. But keep in mind, the X Prize wasn't even funded until halfway through our program. And, in fact, I had an opinion that Peter Diamandis would never get the funds for it. So, we had written him off."

By the time the partnership was finalized a few months into 2001 and Allen provided the funding to Rutan, winning the X Prize had also become a goal of Mojave Aerospace Ventures. "There were two ways for me to recoup my investment," Allen said. "One was the winning of the X Prize, and one was the licensing we'd be able to achieve with a company like Virgin Galactic. Those were the possible

Early Configuration

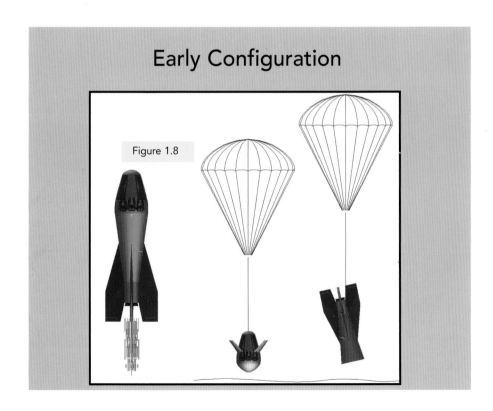

Figure 1.8

Fig. 1.8. After completing the boost, the capsule separated from the booster. It is here where Burt Rutan first applied the idea of a feathered, "carefree" reentry. Small arms pointing upward from the capsule would safely decelerate and steady the capsule during reentry. *Mojave Aerospace Ventures LLC, provided courtesy of Scaled Composites*

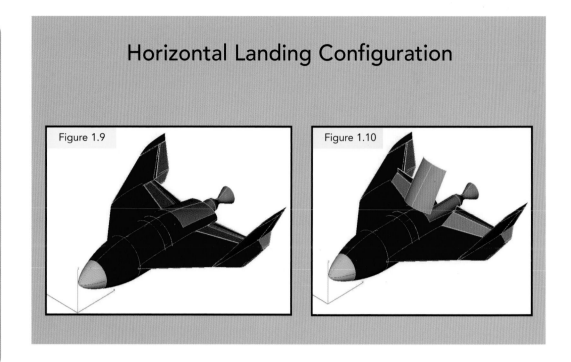

Horizontal Landing Configuration

Figure 1.9

Figure 1.10

Fig. 1.9. Because of Burt Rutan's decades of experience designing aircraft, he decided to abandon the idea of a rocket and parachutes. His next designs focused on winged aircraft that could make horizontal landings on a runway.

Fig. 1.10. Still called a feather, early winged designs used large spoilers and elevons for reentry. Below the speed of sound this configuration worked. However, reentry occurred above Mach 1, and these configurations could not be controlled. *Mojave Aerospace Ventures LLC, provided courtesy of Scaled Composites*

future mechanism of payment back when we were evaluating all this stuff. You didn't necessarily assume you were going to win. And you didn't know what the other competition was like."

Now things started to move full speed ahead.

Tier One

Burt Rutan does not release much information about a newest aircraft while it's under development. In fact, it is usually not until the aircraft is ready to fly that people finally get a glimpse of his newest project. The in-house name of their secret space program was Tier One. This name wasn't uncommon to Scaled Composites. Rutan used "tier one," "tier two," and "tier three" to designate what he described as the "fun factor" of a program. When tier-one programs came around, the company would always bid on them, since they were highly motivational programs for his employees, which helped him retain his skilled staff while being out in the middle of a great big desert. Scaled Composites normally used the aircraft's name as the in-house program name. But this program of course had two vehicles.

"I was making a point to my employees that this was going to be the most fun program that we've ever done and the most important one for us to do," Rutan said.

"That was the original reason behind calling the *SpaceShipOne* program Tier One. Later on when we started to entertain if we should do other manned spacecraft, I just had a feeling that I should make a category for different basic areas of manned spacecraft. Because it seemed like a good breakdown, I defined that if we do programs in the future that send people to Earth orbit, it would be called Tier Two. If we do things that send people outside of Earth orbit to other heavenly bodies like the Moon and Mars, it would be called Tier Three."

The spacecraft and the carrier aircraft needed names, too. The spacecraft was Rutan's Model 316, and the carrier aircraft was Model 318. "Those I assign when I first look at the requirements and

have the first idea of what would be the configuration solution. Those model numbers were assigned years before we had a funded program for building."

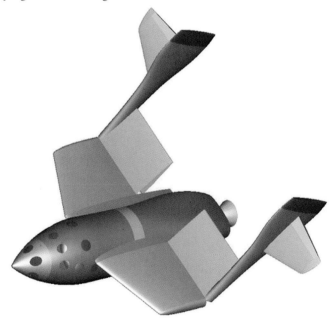

Fig. 1.11. The feather mechanism, the most innovative feature of *SpaceShipOne*, allowed the rear half of the wings and the tail booms to fold upward, which prevented dangerous heat buildup upon reentry but enabled a very stable descent. *Mojave Aerospace Ventures LLC, provided courtesy of Scaled Composites*

Fig. 1.12. Since wind-tunnel testing was not part of the design and development of *SpaceShipOne* because of the large expense, Scaled Composites used a computer method called computational fluid dynamics (CFD) to model the aerodynamics. The photograph shows designer Burt Rutan and aerodynamicist Jim Tighe reviewing a computer analysis of *SpaceShipOne*. Mojave Aerospace Ventures LLC, photograph by Scaled Composites

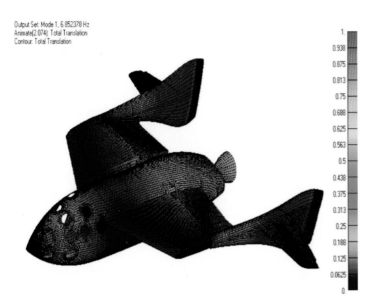

Output Set: Mode 1, 6.852378 Hz
Animate(2.074): Total Translation
Contour: Total Translation

Fig. 1.13. By using computer analysis, Scaled Composites was able to evaluate the aerodynamic characteristics of *SpaceShipOne* even before construction got started. However, it was still necessary to test *SpaceShipOne* in flight to get the complete picture. *Mojave Aerospace Ventures LLC, provided courtesy of Scaled Composites*

SpaceShipTwo and its launch aircraft are Models 339 and 348, respectively.

Rutan added, "We've only flown thirty-nine manned airplanes. Most of the concepts don't get funded and developed."

One example of this is Model 317. This was the design that squeaked between the Tier One spacecraft and its mothership. A new-concept vertical takeoff and landing (VTOL) light aircraft, the tail-sitter would take off like a helicopter but fly conventionally.

White Knight was named by Cory Bird, an employee of Scaled Composites who also flew as a flight engineer in the carrier aircraft on several of the test flights. Bird had made a drawing of a knight in white armor, which Rutan thought was very clever. The drawing ended up being the insignia for *White Knight*, too.

And although Rutan names very few of his airplanes, the spacecraft was an exception. "I wanted to make a point that other manned systems that carried people into space tended to be capsules or spacecraft. But if you read fantasy that kids do, it's always a spaceship. And I thought this is interesting in that there has been a lot of manned spacecraft, capsules, and vehicles, and all these boring names. I felt that this might be the first thing that flies people into space, that you have the moxie to call it a spaceship."

So, since he considered it the first spaceship, he mixed the words around with numbers. "And I thought, I don't have any problem calling this *SpaceShipOne* because I want to build a SpaceShipTwo, and I want to build a SpaceShipThree."

Figure 1.14 shows *SpaceShipOne* mated up to *White Knight* before the public unveiling and even before paint schemes were added.

In order to be able to fly, though, *SpaceShipOne* and *White Knight* each needed a tail number, which is unique identification that every aircraft has similar to a car's license plate, and had to be registered with the FAA. *SpaceShipOne* was given N328KF, which stood for 328,000 feet, the boundary line between Earth's atmosphere and space, while *White Knight* was given N318SL, where the 318 model number and the *SL* stood for spaceship launcher.

"We didn't get our first choices on that," Rutan said. "For example, I wasn't particularly enamored by 328KF. I would have rather had 100KM, 100 kilometers. But it was taken."

Along with the tail number, the type of aircraft had to be identified to the FAA. *White Knight* was pretty straightforward, but *SpaceShipOne* wasn't so clear cut. The FAA felt that commercial launch licensing was required for *SpaceShipOne*. This was a somewhat long and drawn-out process. So as not to delay flight testing, Scaled Composites initially registered *SpaceShipOne* as a glider. This made perfect sense because about half the time during a spaceflight *SpaceShipOne* was a glider. But more importantly, the initial flight testing would be done without a rocket engine. So, by calling it a glider first, Scaled Composites was able to buy some time before having to get their commercial launch license, even though Rutan had no intention of using *SpaceShipOne* commercially.

"When I was out in Mojave for the first-time flight of the X Prize that Mike Melvill flew, it was the first time I had a chance to spend some time with Burt's design team," Poberezny said. "And what was striking was the intelligence that he recruited, the youth and the motivation. In other words, they were hungry and they were motivated to be successful and to make a difference. So, I give Burt a lot of credit. When he saw talent, he brought them in."

Fig. 1.14. Shown before their paint schemes were added, *SpaceShipOne* and *White Knight* share virtually identical cockpits and instruments. But where *White Knight* has jet-engine controls, *SpaceShipOne* has rocket-engine controls. *Mojave Aerospace Ventures LLC, photograph by Scaled Composites*

Rutan gave direction and vision to his staff but still allowed them to exercise their own strengths and abilities. And Rutan would depend on the contributions of the entire team at Scaled Composites to get *SpaceShipOne* into space.

On April 18, 2003, slightly more than two years after receiving Paul Allen's backing, Rutan was ready to go public. *SpaceShipOne* could no longer be hidden away in Scaled Composites' hangar. All the components of Tier One were in place, and *SpaceShipOne* was ready for flight testing. Figure 1.15 shows *SpaceShipOne* after the curtain dropped.

White Knight had been flying since August 2002. It was strange enough looking, yet similar enough to the way-out look of *Proteus* that Scaled Composites didn't worry too much about it occasionally being spotted ahead of time. Both these aircraft have been widely described as looking like giant prehistoric insects or spaceships from *Star Trek*.

It would be a full month, though, before *SpaceShipOne* would take to the air for the first time. However, Rutan had a surprise in store for his guests at the coming-out party. He had *White Knight* do fly-bys for everyone gathered at the flightline in front of the Scaled Composites hangar. Figure 1.16 shows *Proteus*, the original spaceship launcher, and *White Knight*, the new spaceship launcher, flying together, and table 1.1 gives the specifications for *White Knight*.

Aside from the two vehicles, the other important Tier One components were revealed, refer to figure 1.17. The test stand trailer (TST) was a partial mockup of *SpaceShipOne* used to develop the rocket engine. A tanker truck called the mobile nitrous oxide delivery system (MONODS) supplied the nitrous oxide (N_2O) to the oxidizer tank on the TST and in *SpaceShipOne*. And the Scaled Composites unit mobile (SCUM) truck was used for ground control, providing

Table 1.1 *White Knight* Specifications

Wingspan:	82 feet (25 meters)
Wing area:	468 square feet (43.5 square meters)
Fuselage diameter:	60 inches (152 centimeters) for maximum outer diameter
Gross weight:	19,000 pounds (8,620 kilograms) at takeoff with SpaceShipOne
Crew capacity:	one pilot (front seat) and two passengers (back seat)
Crew compartment:	"short-sleeved" pressurized cabin
Engines:	two J-85-GE-5 turbojets with afterburners
Thrust:	7,700 pounds-force (34,000 newtons)
Fuel:	JP-1
Fuel capacity:	6,400 pounds (2,900 kilograms)
Payload capacity:	8,000–9,000 pounds (3,630–4,080 kilograms)
Ceiling:	53,000 feet (16,150 meters)

Fig. 1.15. On April 18, 2003, ten years after Burt Rutan had started to sketch-out designs for a spaceship, the curtain dropped to give the public its very first view of *SpaceShipOne*. *Mojave Aerospace Ventures LLC, photograph by David M. Moore*

Fig. 1.17. Tier One, the space program of Scaled Composites, comprised *SpaceShipOne*, the spacecraft; *White Knight*, the carrier aircraft; test stand trailer (TST), the rocket-engine testing platform; mobile nitrous oxide delivery system (MONODS), the nitrous oxide (N₂O) supply tanker; and Scaled Composites unit mobile (SCUM) truck, a ground-control station. *Mojave Aerospace Ventures LLC, photograph by Scaled Composites*

Fig. 1.16. *Proteus*, flying below *White Knight*, first took flight in 1998 and was originally planned as the carrier aircraft for a single-person rocket. As the spacecraft design evolved into *SpaceShipOne*, the larger *White Knight* was required. *Mojave Aerospace Ventures LLC, photograph by Scaled Composites*

Fig. 1.18. Jeff Johnson, project manager for Mojave Aerospace Ventures, sits at the simulator control desk and monitors the progress of a simulation. One of the most critical components of Tier One was the flight-training simulator. Primarily designed by Pete Siebold, it allowed the test pilots to practice and refine the techniques required to fly the challenging trajectory of *SpaceShipOne*. *Mojave Aerospace Ventures LLC, photograph by David M. Moore*

a telemetry link with the TST during rocket-engine testing or *SpaceShipOne* during a flight.

The last component, shown in figure 1.18, was the flight simulator, which Scaled Composites specially designed for Tier One. It was the first flight simulator Scaled Composites ever built and used for one of their aircraft.

Having a sponsor instead of a customer, building with a robust design approach, performing incremental testing, incorporating as many similarities between *SpaceShipOne* and *White Knight* as possible, and conducting extensive pilot training in the flight simulator, *White Knight*, and Extra 300 aerobatic plane would all prove key factors upon which the success of Tier One would depend.

On November 22, 2001, Steve Bennett's Starchaser team, based in the United Kingdom, became the first competitor to launch a vehicle while displaying an X Prize logo. The single-person *Nova* capsule, unmanned at the time, rode atop the Starchaser 4 booster. The rocket reached a height of 5,541 feet (1,689 meters). *Courtesy of Starchaser Industries*

Chapter 2

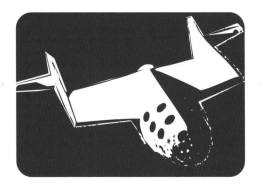

The Ansari X Prize Blasts Off

Prizes and competitions during aviation's infancy sparked what is one of the largest industries today. With that in mind, Peter Diamandis, the founder of the X Prize Foundation, sought to stimulate a similar excitement and interest. But this time the sights were set a little bit higher, 100 kilometers (62.1 miles or 328,000 feet) high to be exact. Here the atmosphere is all but nonexistent, and aerodynamics really don't matter much anymore.

The Ansari X Prize would draw teams competing from Argentina, Canada, England, Israel, Romania, Russia, and the United States. Figure 2.1 and figure 2.3 show some of the many different approaches the teams had in their attempts to snatch the Ansari X Prize.

Yet there was much to overcome. The biggest obstacle was public perception. How could any one of these teams—not governments—accomplish something straight out of the pages of science fiction books?

"When I talked to people during dinner conversation about building a spaceship," said Anousheh Ansari, the title sponsor of the Ansari X Prize, "they completely thought I was a nutcase. They were surprised. Of course now, nobody thinks we're crazy. But back in 2002, you talked about spaceships and building spaceships and no one believed you."

Where there is a will, there is a way to space. But it would not be an easy one. Beyond the idealistic beauty and mystical draw, space is relentlessly unforgiving. There is no pulling off to the shoulder and calling roadside assistance. There is no limping back to the airfield on just one of four remaining engines. Eve the most

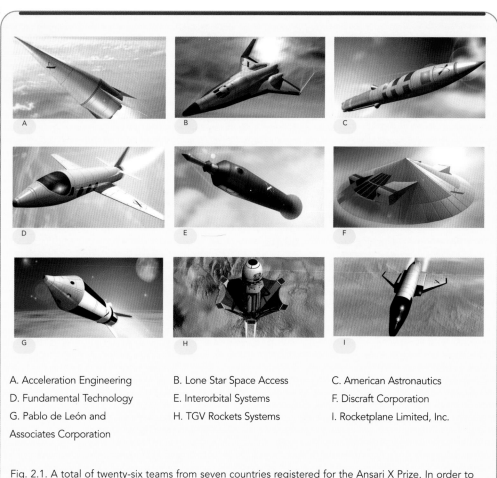

A. Acceleration Engineering
B. Lone Star Space Access
C. American Astronautics
D. Fundamental Technology
E. Interorbital Systems
F. Discraft Corporation
G. Pablo de León and Associates Corporation
H. TGV Rockets Systems
I. Rocketplane Limited, Inc.

Fig. 2.1. A total of twenty-six teams from seven countries registered for the Ansari X Prize. In order to register, teams had to prove they had a well-conceived design and the expertise capable of turning the design into a working spacecraft. Some of the teams were in existence even before the Ansari X Prize was announced, while others formed afterwards. *X PRIZE Foundation*

Fig. 2.2. Spaceflight is risky. Had it not been for the ingenuity of NASA engineers back on Earth and the determination of the crew aboard Apollo 13, the three astronauts would not have survived a crippling explosion that forced them to abort their mission before reaching the surface of the Moon. Their damaged service module is shown. But history shows that aviation during its infancy was just as perilous, if not more so. *NASA–Johnson Space Center*

careful planning cannot completely remove the cold grip of space, as in figure 2.2, where the damaged Apollo 13 service module is shown after the crew narrowly escaped catastrophe during an aborted Moon landing.

Is the price worth it? Each and every day people face risk in their homes and once they step outside. It is familiar risk, though. But it doesn't mean this risk goes away just because people become accustomed to it.

"You cannot have great breakthroughs without risk," insisted Diamandis. "By definition, something that is a true breakthrough, the day before it's a breakthrough, it's a crazy idea. If it is not a crazy idea, then it is not a breakthrough. It's a small, incremental improvement. Computing with silicon instead of vacuum tubes was a crazy idea. So, how do you embrace allowing people to try their crazy ideas?

"How do you allow people to take those risks, people who want to take the risks and not regulate against it? I think space is a very risky business still, and that's okay. I had publicly said that during the course of the X Prize, people may lose there lives. But they are doing it for something they deeply believe in."

Peter Diamandis

Like many people, Peter Diamandis' fascination with space began back when he was a child. But unlike many people, he has not stood idly by waiting for the stars to come to him. His obsession with the point where gravity loses its touch, and the places beyond, firmly took root while he was an aerospace engineering student at Massachusetts Institute of Technology. He had the chance to meet astronauts-in-training back then, but this forced the realization that his chances of becoming an astronaut himself were remote and that even if he did make it as one, he would fly to space maybe twice in a decade. Figure 2.4 shows Diamandis as *SpaceShipOne* made its way to space on October 4, 2004.

The government space programs do work well in specific ways, but very few people will ever get the chance to go up. "That wasn't my vision of spaceflight," Diamandis said. "I wanted to go as a private pioneer in my own ship whenever I wanted to go."

Dennis Tito spent $20 million to fly to the *International Space Station* (*ISS*) aboard a Russian Soyuz in 2001, becoming the first

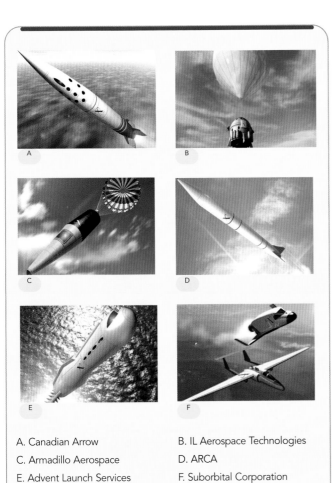

A. Canadian Arrow
B. IL Aerospace Technologies
C. Armadillo Aerospace
D. ARCA
E. Advent Launch Services
F. Suborbital Corporation

Fig. 2.3. The competitors pursued many different approaches, although not every one managed to launch hardware. Concepts were either ground-launched or air-launched, and while most were rockets, many were space planes, with the exception of one flying saucer that would ride upon "blastwave" pulsejets. The air-launched vehicles were either carried or towed by an aircraft or lifted by a giant balloon. The methods of reentry were just as varied. *X PRIZE Foundation*

Fig. 2.4. Peter Diamandis, the founder of the X Prize Foundation, gives the thumbs up as *SpaceShipOne* makes its way up to space during the second Ansari X Prize flight. After reading *The Spirit of St. Louis* by Charles Lindbergh, Diamandis was inspired to create a space prize modeled after the early aviation prizes. *Dan Linehan*

Fig. 2.5. *Atlantis*, *Discovery*, and *Endeavor* are the remaining three operational Space Shuttles. First launched on April 12, 1981, exactly twenty years after Cosmonaut Yuri Gagarin's first-ever spaceflight, the Space Shuttle had been the only U.S. vehicle to carry people into space for twenty-three years prior to the spaceflights of *SpaceShipOne*. Six Space Shuttles were built, although the first Space Shuttle, *Enterprise*, never reached space. In 1986, *Challenger* exploded during liftoff, and in 2003, *Columbia* broke apart during reentry. *Dan Linehan*

Fig. 2.6. Thousands and thousands of space enthusiasts crowded into the high desert of Southern California to watch the spaceflights of *SpaceShipOne* as Mojave Airport transformed into a spaceport. *Mojave Aerospace Ventures LLC, photograph by David M. Moore*

paying space tourist. Diamandis has reported that the cost to fly the Space Shuttle, shown in figure 2.5 preparing to launch to the *ISS*, has ranged from $500 million to $750 million for just one flight, of which propellants make up only 1 percent of that cost. These figures keep the gate to the space frontier shut pretty tight for most people. There just had to be another way.

While flying together over the Hudson River in 1994, Gregg Maryniak, a longtime friend and business partner, wondered when Diamandis would also get his pilot's license. Diamandis had already stopped and started several times. This was unusual, considering Diamandis' deep desire for space. One might expect that for someone with dreams of traveling among the stars, a pilot's license was a good thing to have. But as history continues to remind us, the shortest distance between point A and point B is not necessarily a straight line. Diamandis was far too consumed with what was well beyond where the air is thin.

"Gregg asked me if I had ever read *The Spirit of St. Louis*," Diamandis said. Maryniak had explained that he received the book as a gift, and it helped motivate him to finish his pilot's license. Shortly after their flight, Maryniak gave a copy of *The Spirit of St. Louis* to Diamandis. But if anything, the book proved to sidetrack Diamandis, resulting in the unanticipated consequence of drastically changing not only how people reach space but also who gets to go.

"As I read that book, I had no idea that Lindbergh crossed the Atlantic to win a prize and that nine different teams had spent $400,000 to win the $25,000 prize," Diamandis said. "And by the time I finished reading the book, the whole idea of the X Prize had come to mind."

What Diamandis realized was that a prize could be the catalyst needed for the development of a new breed of spacecraft that could demonstrate the public's desire for commercial spaceflight. "We needed a paradigm shift," Diamandis said. "People had become so stuck in their way of thinking that spaceflight was only for the government—only largest corporations and governments could do this—it could never be done by an individual. This thinking was paralyzing us, and that was what I was trying to change."

When Lindbergh made his famous crossing, the airplane had been in existence for a little more than two decades. It was still a novelty. Some enterprising individuals foresaw the economic advantages of aviation, while others stoked the fanfare and fervor. As a result, hundreds of aviation competitions were established to see who could fly the farthest, the fastest, the highest. It was as much about pushing the limits as it was about drawing boundaries where none had ever existed.

At a time when aviation was in its infancy, prizes and competitions put its growth on afterburners. And during these times, people could look in the mirror and see themselves in the cockpit, goggles drawn and wrapped in a scarf, without having to use too much imagination. Although some of the flyers were wealthy and privileged and others had renown and notoriety, Charles Lindbergh, an airmail pilot, and others like him, proved aviation was in reach of the common person.

Diamandis saw this vision, only with rocket ships and space helmets. His passion was contagious. He energized many talented and dedicated people who joined this march toward space, contributing thousands and thousands of volunteer hours along the way. Figure 2.6 shows the crowds who gathered to share in this vision.

1927: New York to Paris

In 1919, Raymond Orteig created the Orteig Prize for the first non-stop flight across the Atlantic Ocean from New York to Paris or from Paris to New York. Orteig, born in France, owned hotels in New York City. Prizes had been enticing aviators and aircraft makers for a decade now. Newspapers sponsored them because it gave their readers something exciting to read. Businesses sponsored them because they saw financial opportunity.

Aviation technology was not up to the challenge, and Orteig had to extend the deadline of the prize. Come 1926, still no one had claimed the prize. Only one team made an attempt, but they crashed on takeoff.

On May 20, 1927, with only 20 feet (6 meters) to spare, the Ryan NYP *Spirit of St. Louis* cleared the telephone wires a short distance from the edge of the runway at Roosevelt Field on Long Island. Charles Lindbergh, shown in figure 2.7, had just lifted off for his first solo attempt at crossing the Atlantic Ocean. Several failed attempts had already been made by other competitors by now. Nine teams were in the race to win the $25,000 Orteig Prize. Four men had died trying, and two others, setting out together right before Lindbergh, were lost over the Atlantic.

To make the journey, Lindbergh would have to strip the plane down to the bare minimum to maximize the amount of fuel he could carry. Table 2.1 shows the specifications of the *Spirit of St. Louis*. So much of the aircraft was gas tank, by design, that Lindbergh had to use a periscope to see directly ahead of the aircraft because a gas tank in front

Table 2.1 *Spirit of St. Louis* Specifications

Manufacturer:	Ryan Airlines Company
Type:	highly modified M-2
Wingspan:	46 feet (14 meters)
Length:	27 feet 8 inches (8 meters)
Height:	9 feet 10 inches (3 meters)
Empty weight:	2,150 pounds (975 kilograms)
Gross weight:	5,135 pounds (2,330 kilograms)
Engine:	Wright Whirlwind J-5C
Power:	223 horsepower

Fig. 2.7. In 1927, Charles Lindbergh, an unknown airmail pilot, reshaped aviation after crossing the Atlantic Ocean nonstop in an aircraft for the first time, as he flew the *Spirit of St. Louis* from New York to Paris. The solo flight took 33.5 hours to complete and covered 3,610 miles (5,810 kilometers). *X PRIZE Foundation*

Fig. 2.8. The *Spirit of St. Louis* was specially designed by Charles Lindbergh to make the transoceanic flight. Much of the aircraft was a fuel tank, leaving little room for anything else. Lindbergh had to use a periscope to see in front of the airplane, and he elected not to bring a parachute or radio, to save weight. *NASA–Langley Research Center*

of the cockpit blocked the view forward. In a plane that weighed 2,150 pounds (975 kilograms) empty, it carried 451 gallons (1,710 liters) of fuel for a total takeoff weight of 5,135 pounds (2,330 kilograms).

Back in 1927, if you went down in the water, you were gone. There was no satellite tracking, there were no helicopters or airplanes that you could signal. There was no radar. And shipping was nothing like it is today, so rescue from a nearby vessel was highly unlikely. When Lindbergh got behind the controls of that plane and took off, he was all alone with only the vastness of the Atlantic Ocean more than willing to catch him if he fell. And there was less than just a slim chance of him not making it back.

So, the *Spirit of St. Louis*, shown in figure 2.8, didn't have a radio, navigational lights, or gas gauges. Lindbergh didn't even bring a parachute. A radio didn't do any good over the middle of the ocean and, back in those days, was a lot of weight. The same held true for the navigational lights when the wiring was also factored in. Even gas gauges were redundant, since there would not be much he could do about it if the tanks went dry. But the reverse argument could be made. Each of these could help his chance of survival under some specific circumstances. What if a ship was nearby? Lights and a radio could certainly help. What if he was over land? He should be able to make an emergency landing, but there were circumstances where bailing out was not out of the question. Lindbergh had to balance the potential benefit of each safety item with the problems he would potentially face if he ran out of fuel. And that's how he decided.

"He was thinking his way all the way around the problem, though," said Erik Lindbergh, the grandson of Charles Lindbergh. "I think he minimized every possible risk he could except for lack of sleep. And if he had had a good seven hours worth of sleep, he would have really changed his risk factor."

Lindbergh didn't even use a typical leather pilot's seat. Instead, he used a wicker chair. He did, however, equip himself with four sandwiches, two canteens of water, and an inflatable, rubber life raft.

Lindbergh believed that for a multiengine aircraft, there was only a greater risk of an engine failure, even though most of the other competitors were using that type of aircraft. Today, a Boeing 767 flies overseas with only two engines. If one fails, it still has enough power to reach land by either turning around or by continuing on, whichever distance is shorter. That wasn't necessarily the case for the multi-engine aircraft of that time.

"He was doing things like cutting the corners off of his map, which is really a negligible weight," said Erik Lindbergh. "And yet when you look at the competitors, some of them had champagne and croissants on board so they could party when they got there. But they never made if off the ground. So, attention to detail and reducing the risk factors was critical to him surviving the flight."

Charles Lindbergh became an instant international hero on the evening his wheels touched down in Paris. And people's interest in aviation exploded. Charles Lindbergh said, "I was astonished at the effect my successful landing in France had on the nations of the world. To me, it was like a match lighting a bonfire."

Erik Lindbergh said of his grandfather's accomplishment, "Before he flew across the Atlantic, people who flew in airplanes were known as barnstormers and daredevils and flying fools. And after he flew across the Atlantic, people who flew in airplanes were known as pilots and passengers. It truly was a paradigm shift if there ever was one."

As a result of this new popularity, referred to as the Lindbergh boom, in the United States the number of applications for a pilot's license tripled and the number of licensed aircraft quadrupled during 1927. The number of passengers flying aboard U.S. airlines also dramatically increased from 5,782 in 1926 to 173,405 in 1929. Nowadays, the aviation transportation sector is a $300 billion industry.

The X Factor

Now that the idea was hatched, what to name it?

"The letter *X* initially stood for the variable for the person's name that funded the prize, just like the Orteig Prize," Diamandis said. "It worked because $10 million was the number I thought was the right number. I wanted it to be enough money to be of substantial importance to the world, but not so big that it would attract the Lockheeds or Boeings. I didn't want the winner to be a traditional player. I wanted it to be somebody who was going to really work hard on how to do this thing cost effectively and worry about every penny spent."

Finding a title sponsor to put up the prize money proved very difficult, so the X hung around for a lot longer than Diamandis had anticipated. But when the title sponsor did come along, the *X* had already become symbolic. *X* stood for the Roman numeral ten, as in

Fig. 2.9. Initially, the *X* in the X Prize was only a place holder to be replaced when Peter Diamandis found a title sponsor. But gradually it took on its own significance. *X* stood for $10 million, *X* had been used for the early X-planes, and *X* meant mysterious or extreme. So, when a title sponsor did come along, the *X* remained. *X PRIZE Foundation*

the number of millions in the prize. *X* denoted a vehicle of an experimental nature, as with the X-planes. *X* also had the connotation of being extreme or mysterious. "So, after we found the Ansaris," Diamandis said, "we decided to keep it and make it the Ansari X Prize." The logo is shown in figure 2.9.

Rules of the Game

When Burt Rutan rolled out *SpaceShipOne* in April of 2003, he complimented Diamandis, saying that the Ansari X Prize rules had stood the test of time and that it was the brilliant set of rules that allowed the competition to proceed.

After coming up with the concept of the Ansari X Prize in 1994, it took better than a year and a half to nail down the rules. Diamandis stated that the rules were 80 percent of battle. Making them simple, understandable, and bulletproof was an imperative. The rules had to define a precise goal that was very difficult to reach but not completely unattainable. The rules in brief are given in table 2.2.

"I consulted with many of the people who would become teams later on. I reached out to many of the entrepreneurial players in the space community to get their input," Diamandis said.

Table 2.2 Anasari X Prize Rules in Brief

The spacecraft must:

[1] reach a suborbital altitude of 100 kilometers (62.1 miles or 328,000 feet)

[2] carry three people (or one pilot plus the equivalent weight of two other people)

[3] repeat the same flight within two weeks

[4] be designed, built, and launched using only private funding

[5] return safely to Earth with crew unharmed

Initially, the rules were drafted to require the spacecraft to reach an altitude of 100 miles (160 kilometers). By comparison, *Sputnik* orbited above Earth at a maximum height of 588 miles (947 kilometers), while the *International Space Station* typically orbits a little more than 200 miles (320 kilometers) up. But when the reentry characteristics of a spacecraft returning to Earth from 100 miles (160 kilometers) up were analyzed, the heating was determined to be too high. The expense and time to develop an engineering solution for this would have been cost prohibitive to many—if not all—of the teams. An altitude of 100 kilometers (62.1 miles or 328,000 feet) was then selected.

This number was not exactly easy to reach, though. "There was a big debate about what was officially space," Diamandis said. "The U.S. Air Force viewed it at 50 miles [80.5 kilometers], and the Europeans looked at it as 100 kilometers [62.1 miles]. We didn't want the X Prize to be in contention, so we moved it to the higher of the two."

In order for a contestant to claim the Ansari X Prize, it was necessary to verify the altitude that was reached. So, each spacecraft would have to carry a flight recorder, also known as the gold box, provided by the X Prize Foundation to monitor the flight profile. Figure 2.10 illustrates the altitude requirement of the Ansari X Prize.

The next rule to decide was how many people the spacecraft would have to carry. "I didn't want the vehicle to be considered a stunt," explained Diamandis. "I wanted the vehicle winning the X Prize to potently go into revenue service. So, we basically focused on having a vehicle that could fly with a pilot and two paying passengers and required that the vehicle have three seats."

However, three people were not actually required to occupy the spacecraft during the attempts to reach space. The rules state: "The flight vehicle must be built with the capacity (weight and volume) to carry a minimum of three adults of height 6 feet 2 inches (188 centimeters) and weight 198 pounds (90 kilograms) each."

So, right before a launch attempt, three people had to strap into the spacecraft while it was on the ground in order to show that they fit. For each passenger not remaining onboard for the launch, 198 pounds (90 kilograms) of ballast would be added as a replacement.

"And the key rule, which was probably most important, was that the vehicle had to do two flights within two weeks," Diamandis said. "What that meant was that the cost of the second flight was really touch labor and fuel."

This rule was crucial in demonstrating the robustness of the design because it required that 90 percent of the spacecraft's mass, excluding propellant mass, had to be original and could not be

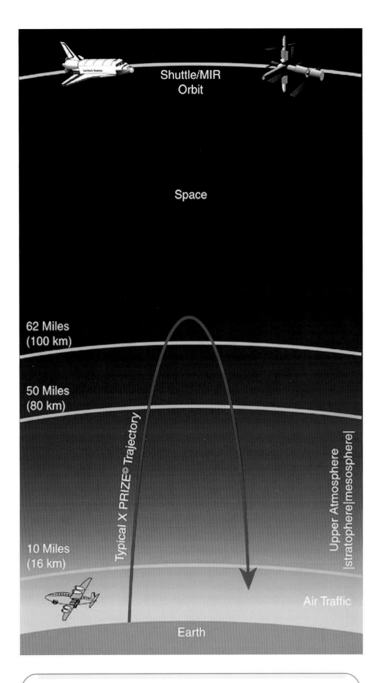

Fig. 2.10. To win the Ansari X Prize, a team had to build a suborbital spacecraft to reach a height of 100 kilometers (62.1 miles or 328,000 feet). This is about a third of the altitude reached by the Space Shuttle and *International Space Station*. It is called suborbital because the altitude is not sufficient for a spacecraft to achieve orbit. *X PRIZE Foundation*

replaced. The spacecraft had to return substantially intact. If after the first flight major components were damaged to the point where they could not be used again, or if too many materials had to be swapped out, then this rule would filter out those designs that were not durable or reusable.

"My mission in the X Prize was to bring about a new generation of privately owned and privately operated spacecraft that can service a marketplace," Diamandis said. Many people, including Diamandis, viewed the space industry, in terms of human spaceflight, as stagnant. The two primary types of vehicles used to escape the confines of gravity remained for decades the Russian Soyuz and the U.S. Space Shuttle, which first flew in 1967 and 1981, respectively. There is a better chance of a person winning the lottery than flying aboard one of these spaceships. And even in the case of a lucky golden ticket, this would not necessarily secure a seat. Since government space agencies operate these vehicles, they have little interest in extending opportunities to the public at large.

This shortcoming of government space agencies was specifically what Diamandis wanted to challenge by requiring that the Ansari X Prize be privately funded. To that end, the following rule was put into place:

> Flight vehicles will have to be privately financed and built. Entrants will be precluded from using a launch vehicle substantially developed under a government contract or grant. Entrants will be prohibited from receiving any direct funding, subsidies, and grants of money, goods, or services from any government (or otherwise tax-supported entity). Entrants will be permitted to utilize government facilities if access to such facilities is generally available to all entrants. Any such goods or services used in connection with the competition must be available to other entrants on similar terms. Entrants will be permitted to utilize subsystems previously developed by a government agency that are currently available on a commercial or equal-access government-surplus basis, or for which manufacturing rights and specifications are available on an equal-access basis.

The competition was not without risk to its participants. However, since the whole idea of the Ansari X Prize was to promote public space travel, one of the more obvious rules was that the crew had to return to Earth safe and sound after each attempt.

The Lindbergh Legacy

Early on, as the X Prize Foundation started to pull together, Diamandis and Bryon K. Lichtenberg, who was one of the cofounders of the organization, met with Erik Lindbergh. Lichtenberg, a two-time shuttle astronaut, was one of the very first people Diamandis spoke to about the X Prize idea and was his partner in Zero Gravity Corporation.

Diamandis had felt it important to have a connection with the Lindbergh family. But the Lindbergh Foundation wasn't initially interested because its focus was on supporting projects that emphasized the balance between technology and a healthy planet. "I think people have sort of lost that dream of space travel after Apollo faded and space flight became routine and boring," said Lindbergh.

But it was hard for Lindbergh's passion not to be stirred up by the ideas of Diamandis. "What really got me was the fact that this could

ignite that kind of inspiration again," said Lindbergh. "And there is a tremendous amount of knowledge that we need for space travel that will translate directly into the quality of life here on Earth, such as environmental technology and closed-loop living systems."

Lichtenberg talked about the view from space aboard *Columbia* in 1983, where he was the very first Space Shuttle payload specialist, and of his time on *Atlantis* nine years later. Lindbergh remembered reading about Frank White and his overview perspective of Earth from space, as well as Jim Lovell being able to cover up Earth with his thumb as he looked out the window of his space capsule.

"These astronauts had an overview perspective that could be tremendously valuable in terms of how we navigate the present so that we can thrive and survive into the future," said Lindbergh. He participated in the unveiling of the X Prize under the St. Louis Arch in 1996, but he would soon be drawn in much deeper.

The Cash Prize

This was all a lot of careful planning, but $10 million of prize money does not just materialize out of thin air. This figure does not even include all the expenses needed to run the competition. All totaled, it was a lot of spacebucks. The early conceptual drawing in figure 2.11 gives an indication of the broad, forward thinking of the ambitious X Prize.

In 1995, Diamandis established the X Prize Foundation in Rockville, Maryland, with the help of Maryniak, Lichtenberg, and Colette M. Bevis. That same year Diamandis met Doug King, the president of the St. Louis Science Center, who offered to help raise $2.5 million if the X Prize Foundation would relocate there. St. Louis embraced Diamandis, and with its aviation heritage, the decision to move was an easy one.

During a fundraiser in St. Louis, a local businessman named Alfred Kerth reminded Diamandis that Charles Lindbergh created the Spirit of St. Louis Organization. This organization was a group of ten business leaders who contributed a total of $25,000 to purchase the aircraft used by Lindbergh to cross the Atlantic Ocean. "And *Spirit of St. Louis*—the airplane—was named after that organization," Diamandis said. "So Kerth said, 'Let's get one hundred people to contribute $25,000 or more from the St. Louis region and call them the New Spirit of St. Louis Organization. It will be the funding mechanism to kick this whole thing off.'"

On May 18, 1996, three days before the anniversary of Lindbergh's historic flight, under the St. Louis Arch, the X Prize was announced. Guests of the ceremony included twenty astronauts; Dan Golden, the administrator of NASA at the time; the Lindbergh family; and Burt Rutan, who on that day made his interest clear. The race was on, and teams had until January 1, 2005, to claim the X Prize.

By 2001, the X Prize was still not fully funded. Bob Weiss, movie producer and vice-chairman of the X Prize Foundation, proposed the idea of a hole-in-one insurance policy to Diamandisrize. With a hole-in-one insurance policy, an insurance company essentially bets against an event happening. This is not uncommon in golf tournaments, where a player can win a car or a great deal of money if he or she makes a hole-in-one on a specific hole on the golf course. If no player makes a hole-in-one, then the insurance company keeps the insurance premiums paid by the tournament organizers and pays nothing out. However, if a player does make a hole-in-one, then the insurance company pays the check.

Fig. 2.11. The vision of Peter Diamandis and the X Prize Foundation was to rekindle the public's interest in space and foster the development of private spacecraft that would open the door to the stars for more than just the very limited number of astronauts from government-sponsored programs. *X PRIZE Foundation*

The X Prize Foundation moved ahead with the insurance idea, but premiums were not inexpensive. "I would have to pay out $50,000 every other month sometimes and a large balloon payment at the end," Diamandis said. "And there were times that I would literally have a week in which to raise $50,000 or I would lose all the premiums I had paid earlier."

After being in existence for six years, the X Prize was much more fragile than most people knew. It was very difficult to raise money to support the day-to-day operations, let alone funding the $10 million prize money.

During the height of Erik Lindbergh's involvement, he had become the vice president of the X Prize Foundation. In 2002, he retraced his grandfather's famous flight on the 75th anniversary of the historic crossing of the Atlantic Ocean by the *Spirit of St. Louis*. Flying a modern Lancair Columbia 300, named the *New Spirit of St. Louis*, Eric Lindbergh flew the same flight path but did so in a little more comfort and safety. He could actually see out the front windshield and did not require the use of a periscope. He averaged 184 miles per hour (296 kilometers per hour), and the flight lasted 19.5 hours compared to 108 miles per hour (174 kilometers per hour) and 33.5 hours for his grandfather's transatlantic flight.

"When I decided to fly across the Atlantic in the Columbia, I did it really to support X Prize," Erik Lindbergh said. "That was the main thrust of it. That was one of many efforts by individual directors that saved X Prize at a specific period in its history."

Almost one million dollars was raised, with a majority going to the X Prize. But that wasn't enough to keep it from ditching before reaching the final destination.

Anousheh Ansari

Anousheh Ansari's fascination with space and the stars began when she was a little girl living in her native country of Iran. At sixteen, she and her family immigrated to the United States. Ansari, shown in figure 2.12, did not speak English, but education was extremely important to her family. She would pick up the language, a bachelor's in electronics and computer engineering, and a master's in electrical engineering on the way to co-founding Telecom Technologies, a multi-million-dollar telecommunications company.

In all this time, her desire for spaceflight never wavered. "Because I didn't become a professional astronaut, I have been looking for other ways," Ansari said. "So, even before meeting Peter Diamandis, I did a lot of looking around on the Web and other places, trying to see what was happening with the space program and if there would be an opportunity for civilians to fly. I had visited the X Prize website and a couple of other websites where they were advertising for tickets for suborbital flights. I did a little bit more research and found out they were basically just doing a lot of conceptual design of these suborbital vehicles to compete in the X Prize. I believed that it would happen soon enough, and probably my first experience or first chance would be on suborbital flight."

In 2001, *Fortune* magazine ran an article about the forty wealthiest people under the age of forty. Ansari made number thirty-three on the list, ahead of Jim Carrey at number thirty-six and Tiger Woods at number forty. But in a sidepiece, Ansari made it clear to the world that space was her number-one goal. There, Ansari had expressed "her desire to board a civilian-carrying, suborbital shuttle."

"I read that like three times," Diamandis said. "So, I convinced myself that it really said suborbital flight."

Fig. 2.12. Captivated by space from her childhood days, Anousheh Ansari never stopped believing that some day she would make it to space. In 2004, the Ansari family was officially named the title sponsor of the X Prize. Two years later, Anousheh Ansari's dream came true. *Prodea Systems, Inc. All rights reserved. Used under permission of Prodea Systems, Inc.*

Diamandis and Lichtenberg immediately contacted Ansari to arrange a meeting. "From the first moment we sat across the table and started to talk about it, Peter had us sold," Anousheh Ansari said, speaking of her and her bother-in-law, Amir Ansari, who had shared the same excitement about space.

Ansari began backing the X Prize in 2002. However, it wasn't until May 2004 that the Ansari family was announced as the title sponsor. "Our sponsorship was absolutely needed for X Prize to succeed," she noted. "At the time we joined the organization, if we had decided not to, I don't know if they would have survived. We felt that we couldn't let that happen. This was too valuable. It was difficult to put together such a good group of people again. The momentum was right. We

couldn't just let it go. And at the same time, the reason we did it was because we love flying to space. And it wasn't like I want to do it just once, and we knew there were millions of people around the world that felt the same way. We wanted to do something to help build an industry so this would become something that would be available, and you can do it again and again and again."

Gathering Momentum

Just as the X Prize Foundation faced challenges to keep the prize going, the individual teams faced similar problems. The biggest of these was funding. It wasn't so much of a technology challenge that the teams had to overcome—the technology to get into space had been around for a long time. The teams were not starting from square one. In fact, with modern materials and computers, a technical leap wouldn't likely be the limiting factor.

Rutan and his team certainly seemed to be in a very good position. "Some viewed him as the frontrunner," Diamandis recalled. "Lessons of history are that sometimes the frontrunner doesn't win. In the Orteig Prize for example, the frontrunner was Admiral Byrd, the first person to fly to the North Pole. He crashed on liftoff. This young upstart, unknown to the rest of the world, Charles Lindbergh, comes along and wins the competition."

A total of twenty-six teams registered for the Ansari X Prize, representing seven different countries, but not every team that applied made it in. "We probably turned away about half the applications we received," explained Diamandis. "We required the teams to really demonstrate to us the seriousness of their team and effort. They had to demonstrate by virtue of the people who were involved, the companies who were involved, and they had to show us the primary concept.

"We had numerous teams apply with antigravity and UFO technology. My answer was simple: 'My office is on the second floor. If you can float up to the second floor, I'm happy to register you.'"

Some of the teams that competed did get vehicles into the air and performed flight tests to various degrees, while some hadn't had the resources necessary to get their programs very far off the ground. Two of the top contenders were Steve Bennett's Starchaser out of England and Brian Feeney's da Vinci Project out of Canada, refer to figure 2.13 and figure 2.14, respectively.

"Steve Bennett was the first person to fly an X Prize vehicle, or a vehicle with X Prize logos on it, called *Nova*, which was the first launch in like thirty years out of the UK," Diamandis said.

Launched on November 22, 2001, *Nova* weighed in at 1,643 pounds (747 kilograms). It was unmanned, but the capsule was designed to fit one person. Starchaser had actually developed and launched rockets beginning in 1993, well before the Ansari X Prize was announced, so it was one of the more established teams.

"I didn't want to apply for the X Prize competition and come across as someone who was ill equipped to deal with it," said Steve Bennett, the team leader of Starchaser. "It took me about a year to get all my ducks in a row. And we made the application, and Peter accepted it no problem."

A bigger vehicle was still needed, since the Ansari X Prize required it to fit three people. "We went through a number of ideas and over the years the design evolved," continued Bennett. "And what happened was it just got simpler and simpler and simpler. So, we

Fig. 2.13. Having launched over sixteen large rockets since 1993, Starchaser was one of the more experienced teams competing for the Ansari X Prize. Even after the Ansari X Prize, Starchaser rocket development continues. In 2007, Starchaser won a study contract from the European Space Agency (ESA) to further investigate reusable launch vehicles for space tourism. *Courtesy of Starchaser Industries*

Fig. 2.14. To avoid the high cost of developing a ground-launched rocket or a carrier aircraft, Brian Feeney's da Vinci Project built the world's largest balloon, capable of holding 3.70 million cubic feet (0.10 million cubic meters) of helium, to lift its *Wild Fire* rocket to a launch altitude of 70,000 feet (21,340 meters). *Brian Feeney, the da Vinci Project*

ended up with a very simple ballistic rocket. We discovered that the easiest and simplest and safest way to do this would be to just launch a ballistic rocket just like they launched Alan Shepard in back in '61. Straight up, straight down, and a capsule that can carry three people. And that's pretty much it."

The rocket had three main components: a booster, which was the Starchaser 5 rocket powered by liquid oxygen (LOX) and kerosene fuel engines; the capsule, which was called *Thunderstar*; and the launch escape system.

Figure 2.15 shows Bennett standing next to the *Nova 2* after piloted drop tests from 10,000 feet (3,050 meters) were conducted.

Bennett admitted, "The biggest challenge was, I guess, raising the finances, because the technology to do this kind of thing has been around since the 1950s, possibly even the 1940s."

So, the team had to be creative. Back in 2000, they had pre-sold two of the seats for when they would first attempt the Ansari X Prize.

Fig. 2.15. Steve Bennett stands next to *Nova 2*, which was drop tested from an altitude over 10,000 feet (3,048 meters) to test the parachute-recovery system. For each test, a pilot manned the capsule, which was then dropped out the back of a cargo plane. *Courtesy of Starchaser Industries*

Bennett said of the two prospective passengers, "They wanted to basically support the project. And they wanted to fly on the first flight. We got three seats in the capsule. The only condition they made was, 'Here's the money, Steve. We'll give you the money. We'll give it to you up front, and we're not even going to come back to you. We don't care whether it takes a year or ten years. You tell us when it's ready. We're not going to hassle you. There is only one condition.' And the one condition was the third seat had to be occupied by me. Okay. So, they knew I wasn't a nutcase. They knew that I wanted to do this project and that I wanted to come home to my family."

The competition really began to heat up midway through 2004, when two teams each secretly notified the X Prize Foundation that they were going for it. Diamantis recalled, "People have to give both confidential notice within 120 days and public notice within 60 days of their attempt to fly. So, we had gotten confidential notice of Rutan's flight date, and then a few weeks later we got confidential notice from Brian Feeney that the da Vinci Project was going to attempt a flight.

"We thought for a moment we might have to mount flight competitions in two separate nations, and funding that and doing a good job of judging was going to be a challenge for us."

Feeney, who had an aerospace company in the 1980's that did life-support systems, read an article about the announcement of the Ansari X Prize while living in Hong Kong. "That was the catalyst for me," he said. "I stopped what I was doing right away at the time."

Feeney had constantly looked for opportunities that would take him to space. He rallied one of the largest volunteer efforts for a technology project in Canadian history while using a very unconventional approach to reach space. At first he looked into developing a carrier aircraft like Scaled Composites had done. "We wanted to do that even before we knew what they were doing. But the cost to do that was just prohibitive. We knew we'd be challenged just to get the money to build the spaceship itself," Feeney said.

"A ground-launched vehicle required about four times as much energy, thrust, everything else, compared to an air-launched vehicle, whether it was balloon, as in our case, or an aircraft or whatever

Fig. 2.16. Teams had to give a 120-day confidential notice to the X Prize Foundation that they were going to make an attempt at the Ansari X Prize. As the deadline loomed, Scaled Composites gave its secret notification, and a few weeks later, the da Vinci Project also gave notice. The da Vinci Project finally got a big chunk of funding from GoldenPalace.com, but it proved to be too late in the game. *Brian Feeney, the da Vinci Project*

means of getting to that altitude. At the end of the day, I thought the balloon-launch system was a good compromise with the much higher cost of developing a ground-based launch rocket."

A rocket called *Wild Fire*, pointing up at 75 degrees, would be tethered to the world's largest reusable helium balloon. Made of polyethylene, the balloon, measuring in at 150 feet (46 meters) in diameter and 200 feet (61 meters) in length, could contain 3.70 million cubic feet (0.10 million cubic meters) of helium. It would only be 15 percent full at liftoff but would expand as it ascended. The da Vinci Project had successfully tested a scaled-down version of their launch balloon before constructing the final launch balloon.

Figure 2.16 shows the *Wild Fire* rocket at only 80 percent complete. At an altitude of 70,000 feet (21,340 meters), which is higher than any launch aircraft could fly, *Wild Fire* would detach and use a hybrid fuel rocket engine to soar into space.

"I never felt in the long run that it was an ideal commercial proposition, because the balloon is subject to weather and a multiplicity of things. But it is a cost-effective way for a short program to demonstrate viable technology," Feeney said.

Threads of Rocketry

Robert Goddard built and launched the very first liquid-fuel rocket in 1926, shown in figure 2.17, and is now considered the father of American rocketry. Yet when he first published theories about reaching the Moon using a rocket, he drew ridicule, so much so that he fled to the remoteness of Roswell, New Mexico, to continue his pioneering research. Ironically, it was prewar Germany that seemed to take his work more seriously than did his own country.

But Goddard still had a few pretty important believers in his work. Charles Lindbergh had recognized that wings and propellers could

Fig. 2.17. Robert Goddard was a pioneer in rocketry. In this photo from 1926, he stands next to his invention, the very first liquid-fueled rocket. His breakthroughs were, unfortunately, not enough to shield him from relentless ridicule when he asserted that a large enough rocket could reach the Moon. *NASA–Marshall Space Flight Center*

Fig. 2.18. Charles Lindbergh (right) understood early on that airplanes had limitations. He was fascinated by rocketry because he knew this would be the next step. A kindred spirit, Lindbergh helped Robert Goddard (center) to get funding from the Guggenheim family. This photograph from 1935 shows Harry F. Guggenheim (left) visiting for a rocket launch. *NASA-Headquarters*

Fig. 2.19. The first man-made object to reach the edge of space, the V-2, was designed and built by Germany during World War II. Thousands and thousands of these ballistic missiles rained down on cities, with Antwerp and London being the targets of the vast majority. After the war, this vengeance or terror weapon became the foundation of the intercontinental ballistic missile (ICBM) and space programs of the United States and Soviet Union. *NASA–Marshall Space Flight Center*

carry an aircraft only so high. Because of this, Lindbergh was very interested in rocketry and in Goddard's work. Lindbergh was able to help secure funding for Goddard from the wealthy Guggenheim family. Figure 2.18 shows Lindbergh together with Goddard and his benefactor for the launch of one of Goddard's more advanced rockets.

Right after World War II, when the German rocket scientists who worked on the V-2 rockets left for the United States or Russia, the United States asked Lindbergh to go to Germany and assess what was left of the V-2 program. Figure 2.19 shows a V-2. It was not only a weapon of war, but it also was the foundation upon which the U.S. space program was built. Even an Ansari X Prize team, Canada's Red Arrow, used it as the basis for their spacecraft.

So, early on, Charles Lindbergh was exposed to the work of Wernher von Braun. And it was von Braun who helped the United States to reach the surface of the Moon. It is hard not to see all the connections and parallels between this and the present.

SpaceShipOne may look like an airplane, but it is actually a combination of a missile launching from an aircraft, a spaceship maneuvering above the atmosphere, and a glider drifting down to Earth. *Mojave Aerospace Ventures LLC, photograph by David M. Moore*

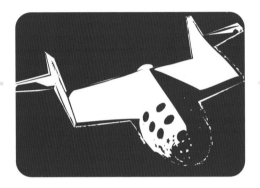

Chapter 3

Flying the Spaceflight Mission

The objective was clear: a suborbital spaceflight above 328,000 feet (100,000 meters). But this mission was far from being straightforward. *SpaceShipOne* was groundbreaking. And although getting off the ground wasn't too difficult, being able to return safely to the ground with the spacecraft intact required facing some pretty tough challenges. Three test pilots, Brian Binnie, Mike Melvill, and Pete Siebold, each flew *SpaceShipOne* during flight testing and handled the curves, and sometimes spirals, thrown their way. "Well, it is kind of a scary little thing to fly," said Mike Melvill of the spacecraft he piloted for the first six flights and four subsequent others.

Figure 3.1 shows *SpaceShipOne* and *White Knight* waiting for launch in the early morning sunlight.

A carefully designed vehicle flown by a highly skilled pilot could tame the perilous environment and extreme forces, but it wasn't easy. *SpaceShipOne* flew a total of seventeen times during its pursuit of space and the Ansari X Prize: three captive-carry flights, eight glide flights, three powered fights within the atmosphere, and three suborbital spaceflights. The combined flight time for these was 4 hours, 11 minutes, and 4 seconds, while the total burn duration was 5 minutes and 47 seconds. The spaceflight profile flown by *SpaceShipOne* is shown in figure 3.2.

White Knight would lift off the runway at Mojave Airport with *SpaceShipOne* slung below. And as the pilot anxiously awaited separation, all he could do was

Fig. 3.1. With *SpaceShipOne*'s oxidizer tank filled the day before a launch and a fully fueled CTN (case/throat/nozzle) installed, *SpaceShipOne* and *White Knight* undergo preflight preparations while illuminated by the early morning Sun. *Mojave Aerospace Ventures LLC, photograph by Scaled Composites*

hang on until launch altitude. Upon separation, *SpaceShipOne*'s rocket engine fired and boosted the pilot several times faster than the speed of sound. Since the rocket engine's fuel was rubber, it gave a whole new meaning to *burning rubber*. But when the rocket engine shut down, after burning 76–84 seconds, the spacecraft was still only half the distance to apogee and had to coast the remaining way through the ever-thinning atmosphere.

On the way back down, *SpaceShipOne*'s feather ensured a safe reentry back into Earth's atmosphere. When the craft was no longer falling at supersonic speeds, the feather retracted. *SpaceShipOne*, now a glider, descended to Mojave Airport for a horizontal landing, just like an ordinary airplane. Table 3.1 shows the mission profile for each of *SpaceShipOne*'s suborbital spaceflights.

For a spaceflight, the duration of the entire mission was only 1.6 hours. This was the time it took *White Knight* to take off, release

Table 3.1 *SpaceShipOne's* Suborbital Spaceflight Mission Profile

1. Liftoff of *SpaceShipOne* mated to *White Knight*
2. Captive-carry to launch altitude
3. *SpaceShipOne* separation from *White Knight*
4. Supersonic boost to space
5. Coast to apogee
6. Freefall from apogee
7. Supersonic reentry into the atmosphere
8. Descent with feather still up
9. Gliding descent back to runway
10. Horizontal landing

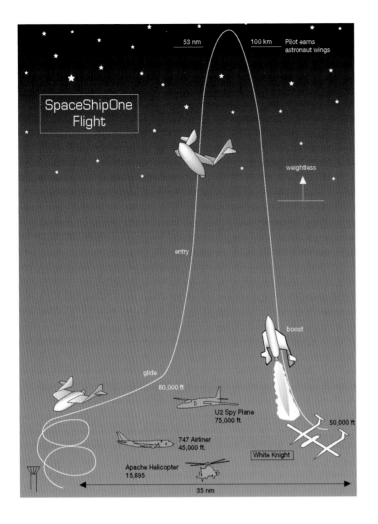

Fig. 3.2. After *White Knight* lifted off with *SpaceShipOne* mounted below, it spiraled up to a launch altitude of about 47,000 feet (14,330 meters), where *SpaceShipOne* then separated, ignited its rocket engine, and boosted to an apogee above 100 kilometers (62.1 miles or 328,000 feet). *SpaceShipOne* reentered the atmosphere with its feather deployed, then reconfigured into a glider for landing at Mojave. *Mojave Aerospace Ventures LLC, provided courtesy of Scaled Composites*

SpaceShipOne, and then land. However, the spacecraft flew for about 24 minutes only, making it back from space and touching down even before *White Knight* did.

Liftoff to Separation

SpaceShipOne, with its landing gear retracted, was rolled on a dolly underneath *White Knight* and then raised using a hand crank. The top of *SpaceShipOne* attached to a two-point pylon that was mounted on the belly of *White Knight*. Two hooks inside the pylon, fore and aft, clamped onto *SpaceShipOne*, securing the vehicles together, as shown in figure 3.3. Heating ducts from *White Knight* to *SpaceShipOne* also ran through the pylon. The wings of *SpaceShipOne* were further secured by braces running down from *White Knight*. The clearance between the two was only a meager 1 foot.

The optimum time to launch from Mojave Airport was at daybreak. Figure 3.4 shows a predawn preflight briefing, which reviewed the mission readiness and addressed any last-minute concerns. After *White Knight*, carrying *SpaceShipOne*, rolled out of the Scaled

Fig. 3.3. *SpaceShipOne* attached to *White Knight* at several points. A pylon underneath *White Knight* housed two hooks that fit into two rings on top of *SpaceShipOne*. A brace between each set of wings also helped to stabilize *SpaceShipOne* and keep it from swaying back and forth. *Mojave Aerospace Ventures LLC, photograph by David M. Moore*

Fig. 3.4. Early in the morning before each test flight, a preflight briefing was held so that the test pilots, ground crew, and Mission Control could review the details of the flight plan and receive updates on the status of the vehicles and the weather conditions. *Mojave Aerospace Ventures LLC, photograph by Scaled Composites*

Fig. 3.6. Inside Mission Control, the flight director, Doug Shane (seated front row center), was the point person on the ground who communicated directly to the pilots aboard *SpaceShipOne* and *White Knight* during all phases of the mission. The Mission Control staff monitored all aspects of *SpaceShipOne* based on the instrument and video data transmitted by the craft. *Mojave Aerospace Ventures LLC, photograph by Scaled Composites*

Fig. 3.5. *White Knight* and *SpaceShipOne* taxi to Mojave Airport's Runway 30. Located less than eighty miles northeast of Los Angeles, Mojave Airport became the first inland spaceport and the first public spaceport licensed by the FAA. *Mojave Aerospace Ventures LLC, photograph by Scaled Composites*

Composites hangar and final preparations were completed, the two vehicles taxied out to the runway. Figure 3.5 shows *White Knight* and *SpaceShipOne* getting ready for takeoff. Once in the air, *White Knight* headed 40 miles (64 kilometers) to a release box, an area that had been designated by the FAA.

Early in the program, the target for the separation altitude was 50,000 feet (15,240 meters). However, both the characteristics of the air, more specifically the density, and the performance of *White Knight*, or lack thereof, favored a separation altitude closer to 47,000 feet (14,330 meters). For this first stage of the ascent, *White Knight* spiraled up at a rate of about 700 feet per minute (210 meters per minute). And by the time the vehicles reached 45,000 feet (13,720 meters), the vehicles were above about 85 percent of the atmosphere.

"We fly up there on *White Knight*, and that is a really long period of time. It is close to an hour to get up there," Melvill said.

"I really hated the ride up there," he continued. "No one wants to talk to you. They think you need to sit there and concentrate on what you are about to do. I really would have liked someone to distract me and have a conversation about something else, because an hour is a long time to sit there and worry about what's going to happen.

"You start getting close to the drop zone and close to the altitude. Then you go through a pretty extensive checklist setting the airplane up. You trim it 10 degrees nose up, so that when it drops off, it holds its nose up."

Fig. 3.7. Tucked underneath *White Knight*, *SpaceShipOne* is carried to a launch altitude of 47,000 feet (14,330 meters). By using a carrier aircraft, Scaled Composites could lift *SpaceShipOne* above about 85 percent of Earth's atmosphere, allowing *SpaceShipOne* to be smaller and thus cheaper to build. Ironically, that launch altitude provided an additional measure of safety in the event of a rocket engine malfunction, since there would be much more time to address a potential problem compared to a ground launch. *Jim Koepnick/Experimental Aircraft Association*

Watching for the slightest abnormality, Mission Control carefully monitored the flight data and video transmitted by *SpaceShipOne* during the entire flight, as figure 3.6 shows.

It was possible to rotate the entire horizontal stabilizer on each of the tail booms. Adjusting it with 10 degrees nose-up trim would help ensure that *SpaceShipOne* didn't go into a dive once it detached from *White Knight*. This setting was also necessary to help force the nose up once the rocket engine fired, enabling *SpaceShipOne* to move from horizontal flight to nearly vertical flight.

Figure 3.7 shows a close-up of *SpaceShipOne* during the captive-carry.

The procedure for the airborne launch of the rocket-powered second stage from the carrier aircraft was pretty simple. Even before the 1920s, flying vehicles had been dropped by larger flying vehicles. And NASA extensively used motherships for dropping X-planes and even for flight testing the Space Shuttle (refer to figure 3.8 and figure 3.9, respectively).

White Knight had to remain steady with its wings level. The pilot in *SpaceShipOne* began the sequence by arming part of his release system. In the cockpit of *White Knight*, a yellow light then came on. *White Knight* would arm next, giving another yellow light. At that point, the release handle inside *White Knight* became "hot." A crew member in the backseat pulled it to retract the hooks. *White Knight* shook with a big bang as the spring-loaded hooks came to a sudden stop. *SpaceShipOne* dropped away as *White Knight* surged upward.

"There is an instant feel of you climbing," said Pete Siebold about flying *White Knight* during this separation. "You lost almost half your weight. You were at 1 g and it jumped up to 1.6 g's the second you pulled the release handle."

Siebold flew *SpaceShipOne* three times, but all the test pilots pulled double duty when it came to flying. During flight tests with *SpaceShipOne*, one of the other *SpaceShipOne* pilots always flew *White Knight*.

Fig. 3.8. Carried to a height of 45,000 feet (13,720 meters) by a NASA B-52, the North American X-15 dropped from the wing at a speed of 500 miles per hour (800 kilometers per hour). The X-15 would typically carry out one of two missions: high-speed hypersonic (faster than Mach 5) test flights or high-altitude test flights. *NASA–Dryden Flight Research Center*

Fig. 3.9. The Space Shuttle *Enterprise* launches from the top of a NASA 747 Shuttle Carrier Aircraft (SCA). During the test flight, *Enterprise* glided back down to Dryden Flight Research Center, so aerodynamic and control characteristics could be evaluated prior to an actual spaceflight. *NASA–Dryden Flight Research Center*

For safe separation from *White Knight*, as shown in figure 3.10, the pilot in *SpaceShipOne* pushed forward on the control stick. This briefly counteracted the trim setting on the horizontal stabilizer. So, instead of wanting to pitch up, *SpaceShipOne* dipped down to avoid re-contacting *White Knight*.

Gliding for about 10 seconds, *SpaceShipOne* grew quiet, since it was now out of earshot of *White Knight*'s engines. The pilot made some quick final checks with Mission Control.

"You have to get the motor started as soon as you possibly can because you are just teetering on the edge of a stall," Melvill added. "You've got all that rubber fuel at the back of the airplane." The center of gravity for *SpaceShipOne* was very far back. If its nose pointed up too high, without the rocket engine going, a stall would develop that would cause the spaceship to lose aerodynamic control. A pilot could recover from this, but *SpaceShipOne* would have lost a good deal of altitude in the process.

Boost and Apogee

Clear of *White Knight*, the pilot armed the rocket engine, flicked the ignition switch, put his head back against the seat, and grabbed the control stick with both hands.

"When you light that rocket motor off, everything literally starts with a bang. There is so much energy associated with that rocket motor. It is like a tsunami sweeps through the cabin and literally takes you away," said Brian Binnie, the pilot who flew the first and last powered flights of *SpaceShipOne*.

"You really have nothing in your background or DNA to tell you that what is happening to you is good. You have no basis. Three or four seconds will go by, and you go, 'Ah, I'm not dead. Therefore, it must be going as they told me it was going to go.'"

The roar from the rocket engine was extremely loud. "The noise is certainly worse right at ignition when you have little forward speed," explained Binnie. "But as soon as you are supersonic, you are going faster then any sound that the rocket motor makes. It doesn't really penetrate the cabin a lot. What you hear in the cabin is all the gurgling off the main oxidizer tank that's right on the other side of the bulkhead from which you're located. So, you hear that. There is a certain amount of wind noise over the vehicle. A good helmet with reasonable ear protection let the radio come across just fine."

The force pinning the pilot to his seat just added to the flood of sensations. "The acceleration is fierce. It's abrupt. It's sudden. It's a big slap in the back and whoosh off you go. It is a very dynamic environment, and you are very much holding on for dear life," Binnie said.

"As it develops, your body readily adjusts to the g's that you are experiencing. They are not that high during boost. They are between 3 or 4."

Having been a former navy fighter pilot, Binnie had some experience with some pretty fast starts. The initial kick from a catapult off the deck of an aircraft carrier had some similarity. After 2–2.5 seconds, the acceleration from the catapult was over, but after the same amount of time, *SpaceShipOne* would still be going and going.

"A catapult shot takes you from 0 to about 150 miles per hour [240 kilometers per hour] in 2 to 2.5 seconds. If you continue that acceleration rate, which is kind of what the spaceship is doing, you would then go from 150 to 300 miles per hour [480 kilometers per hour] in another 2.5 seconds or the 5-second mark. And at the 8-second mark, you'd be doing not 300 miles per hour but 600 miles per hour [970 kilometers per hour]. And by the 10-second mark, you'd be supersonic."

The pilot held the control stick with both hands because, as *SpaceShipOne* moved faster and faster, the forces from the outside air pushed harder and harder against the vehicle and its controls. The pilot had to make sure the nose was coming up right away. Otherwise there was a danger of overspeeding *SpaceShipOne* and breaking it apart. The "never-exceed" speed was around 260 knots equivalent airspeed (KEAS). A knot is a nautical mile per hour, which is a little faster than a mile per hour. Equivalent airspeed is a measure of how fast a vehicle feels it's going in terms of the air pressure pushing against it. So, this value may seem very low, but *SpaceShipOne* started out already operating above 85 percent of the atmosphere. The air density was very low to begin with and didn't exert as much air pressure as if the vehicle were flying at a lower altitude where the air density was higher.

Figure 3.11 shows *SpaceShipOne* during the initial pull up, also called "turning the corner" or the "gamma turn." With the feather locked down tight, it was critical for the pilot to keep the wings level during this phase.

"Because our wings are level, that turn results in pointing nose up," Pete Siebold said. "So, if any portion of the time you roll to a

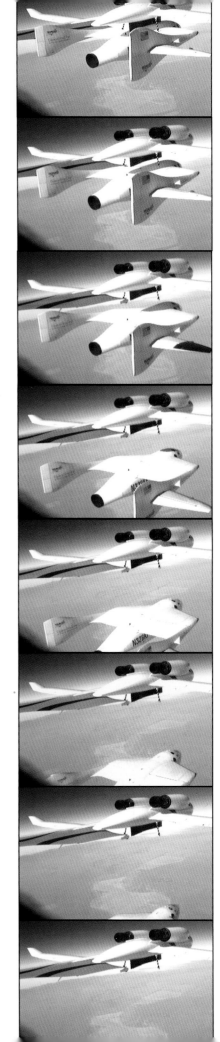

Fig. 3.10. After an arming sequence was completed, the flight engineer in the back seat of *White Knight* pulled the lever that disengaged the two hooks clamped down onto *SpaceShipOne*. The hooks snapped back with a bang, and *SpaceShipOne* fell free. *Mojave Aerospace Ventures LLC, video capture provided courtesy of Discovery Channel and Vulcan Productions, Inc.*

47

Fig. 3.11. Once the rocket engine ignited, the test pilot immediately had to begin to pull the nose of *SpaceShipOne* up in a maneuver called "turning the corner." By the time the turn upward was complete, *SpaceShipOne* was already traveling at supersonic speeds. *Mojave Aerospace Ventures LLC, video capture provided courtesy of Discovery Channel and Vulcan Productions, Inc.*

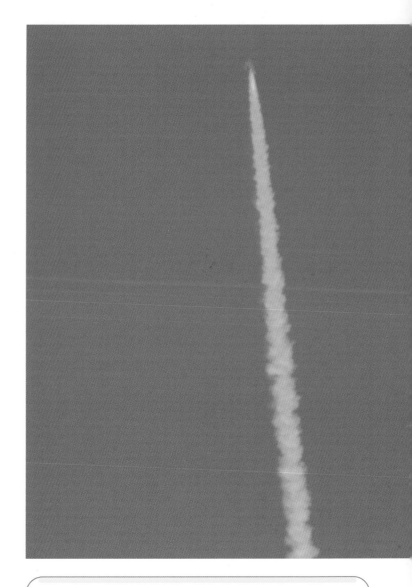

Fig. 3.12. Traveling close to vertical, *SpaceShipOne*'s hybrid rocket engine burned 76–84 seconds during a spaceflight. At rocket engine shutdown, *SpaceShipOne* was a little more than half the way up to apogee. *Mojave Aerospace Ventures LLC, photograph by Scaled Composites*

non–wings level attitude, instead of going up, you are going to turn to a different heading and not go up. A significant amount of the energy is spent doing that initial turn.

"If there was any time spent non–wings level during this turn, you ran the risk of not making your ultimate goal of 100 kilometers [62.1 miles] at the end of the flight."

A number of things caused the wing to wander back and forth. Asymmetries in the thrust from the rocket engine and asymmetries that resulted from supersonic shockwaves wanted to knock *SpaceShipOne* off course.

Melvill added, "And there are wind shears that exceed 100 miles per hour [160 kilometers per hour] going in different directions. So as you go up, it will blow you this way and then it blow you that way. You are not there for long, so you don't get massive changes, but you are constantly correcting."

By using an avionics system called the Tier One navigation unit (TONU), the pilot could ensure that *SpaceShipOne* was wings level and had the proper pitch rate, which was a measure of how fast the nose was rising. For the first 5 seconds, the pilot used the control stick and the rudder pedals to keep the wings level. "You can feel the forces start to build quite rapidly," Binnie said. "And now your thinking is, 'Okay, I'm going to keep fighting this thing physically until about 8 or 9 seconds, and then I'm going to transition over to controlling the vehicle with electric trims.'" To do this, the pilot took his left hand off the stick and reached for the rudder trim controller, a big black knob that looked liked a turtle shell.

Binnie explained, "As you are going through transonics, where the vehicle is shocking up asymmetrically, it is still rocking back and forth or more like whipping back and forth. Your job is to try to filter out the oscillatory motion and chase down with the turtle any longer-term disturbance that is driving the nose of the vehicle off trajectory.

"You have high-rate motions but you have a low-rate controller. So, things don't happen as quickly as you'd like them to. You put in an adjustment, and you're going to have to be patient to sort of see it take effect. And that is not the easiest thing in the world to do because you just had your brains scrambled. And everything about you is on high alert, and now you have to be patient and wait for the thing to respond. If you don't, it is easy to over-control it, and you can get yourself into even more trouble."

As the pilot was finishing the initial pull up, *SpaceShipOne* passed the rough transonic transition from subsonic to supersonic.

"You settle in around the 10- to 15-second mark and look out the window. And appreciate that you are no longer horizontal. The nose will appear vertical, but it is not quite there yet," Binnie said.

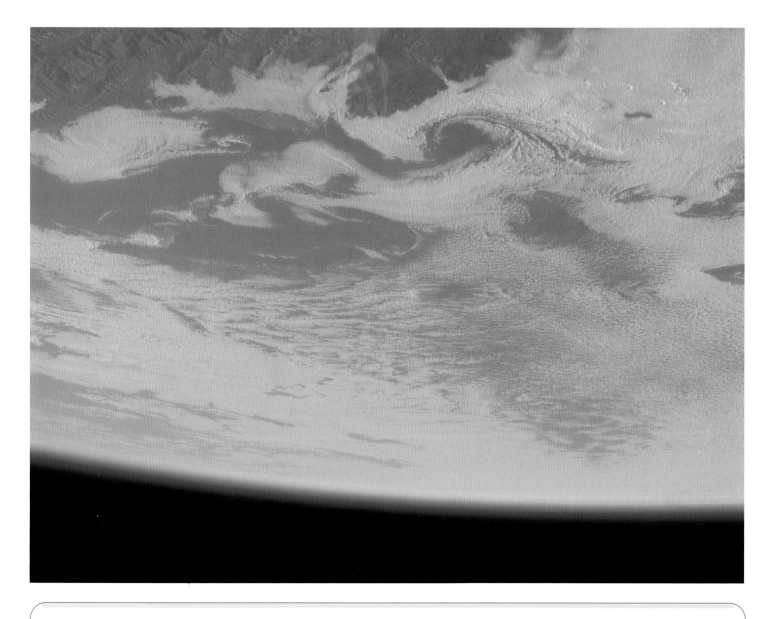

Fig. 3.13. Photographed from inside the cockpit of *SpaceShipOne* by Brian Binnie near the apogee of 367,500 feet (112,000 meters), the Channel Islands and the Pacific Coast peek through the cloud cover as black sky shrouds Earth. *Mojave Aerospace Ventures LLC, photograph by Scaled Composites*

Figure 3.12 shows *SpaceShipOne* and its contrail during the ascent to space.

As *SpaceShipOne* continued to ascend and slowly move its nose closer to vertical, the use of the control stick came back. The pilot could again use the mechanical controls to fly, even though it traveled much faster than Mach 1 and was still gaining speed. The air density was too low to create much opposing force but still high enough for aerodynamics to work.

At about the 1-minute mark, the rocket engine went through a liquid-to-gas transition. "This is kind of a wake-up call that you are getting near the end of the boost phase of flight," Binnie said. "The vehicle shakes and shudders some more. And then the rocket motor valve that you are sitting not too far from has some unusual acoustics associated with it. It sounds like riding along with a possessed cat. It kind of screeches and howls and complains."

SpaceShipOne reached a maximum speed of Mach 3.09, or 2,186 miles per hour (3,518 kilometers per hour). This occurred just before burnout while it was still accelerating. But the atmosphere was very thin at this point, so the airspeed was only about 40 knots equivalent airspeed.

The highest altitude for rocket-engine shutdown occurred at 213,000 feet (64,920 meters). The burn lasted 84 seconds. Unlike shutdown inside the thick atmosphere, where thrust no longer kept the pilot pinned to the seat and the deceleration force flung him forward in his seat, shutdown at high altitudes was tame because of the thin atmosphere and because the rocket engine's thrust had tapered off due to the longer burn duration. *SpaceShipOne* now coasted upward.

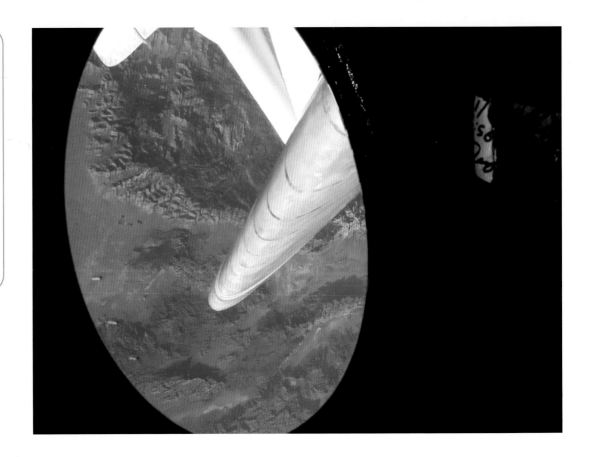

Fig. 3.14. During the 3.5 minutes of weightlessness, the test pilots were able to take photographs from inside *SpaceShipOne*'s cockpit. In this photograph, the red-colored thermal protection on the leading edge of the right wing can be seen through one of the porthole windows. *Mojave Aerospace Ventures LLC, photograph by Scaled Composites*

"Three wonderful things happen," recalled Binnie. "The noise goes away. The shaking, the shuddering, the vibration go away, and you become instantly weightless. And the weightlessness is just a profoundly exciting and pleasurable experience."

"Then of course there is that view. You've seen it on the cover of magazines and things like that. But when you see it for yourself, it is really breathtaking. The eye is so much more dynamic than a video or a camera. It is yours for the enjoyment."

As *SpaceShipOne* rocketed upward, light scattered less and less in the dwindling atmosphere. "You can already see the blue skies turning a much darker, deeper shade of blue and as you continue to watch that, it will deepen and darken and kind of go purplish and then to black," said Binnie.

Figure 3.13 shows the black sky surrounding Earth in a photograph taken by Brian Binnie aboard *SpaceShipOne*. So where are the stars in the black sky? Above the atmosphere there should be stars galore. Well, they are there. However, a camera cannot catch the stars. Earth is much too bright. To see the stars in a photograph, a much longer exposure time is needed, but then the features of Earth would be totally washed out. This phenomenon can also be observed in the famous "Earthrise" photograph taken by the crew of Apollo 8 as they circled the Moon, where Earth is seen rising above the surface of the Moon and no stars can be seen in the background.

Following a ballistic arc, the unpowered spacecraft continued to climb, coasting up while the atmosphere dwindled away. To win the Ansari X Prize, it was necessary to reach an altitude of 328,000 feet (100,000 meters). So, as *SpaceShipOne* raced toward this height, the pilot enjoyed the effects of zero-g. Without the atmosphere, the drag caused by the air resistance was no longer a factor causing *SpaceShipOne* to decelerate. Gravity still had a hold on *SpaceShipOne*, however, and the spacecraft was not traveling at a high enough velocity to escape the pull of Earth.

Having taken an hour to reach launch altitude and separation, the maximum altitude when *SpaceShipOne* reached the top of its climb, or its apogee, occurred about 3 minutes after the rocket engine initially fired off. *SpaceShipOne* stopped moving up at this point and began to free fall back to Earth. The pilot experienced weightless conditions for approximately 3.5 minutes, which he started to feel once the rocket engine shut down. For comparison, the Space Shuttle took 8.5 minutes to go from its launch pad to its orbital altitude of around 200 miles (320 kilometers).

Shot from space, the photograph in figure 3.14 gives a glimpse from *SpaceShipOne*'s window of the leading edge of its wing high above desert mountains.

While outside the atmosphere, *SpaceShipOne* could not use its rudders or elevons to control movement whether it was ascending or descending. Since space is a vacuum, there was no air to provide the lift that these control surfaces required to change the spacecraft's course. This is the same problem faced by the Space Shuttle, as well as other spacecraft and satellites. Even astronauts during extravehicular activity (EVA), floating outside the *International Space Station*, need a way to steer themselves around. To maneuver in space, they accomplished this by shooting little jets of gas in a direction opposite to that of the intended motion. So, for example, if a spacecraft needed to move to the right, it shot a puff of gas to the left.

SpaceShipOne was no different. In the airless environment, the pilot had to use the reaction control system (RCS) to maneuver *SpaceShipOne*.

Fig. 3.15. In an interview with Ed Bradley for *60 Minutes* after the Ansari X Prize spaceflights, Burt Rutan shows the feather configuration and describes how it allowed the safe return of *SpaceShipOne* to Earth's atmosphere. *Mojave Aerospace Ventures LLC, photograph by Scaled Composites*

After he switched the RCS on, he used the control stick and rudder pedals to control thrusters mounted on the fuselage and wings.

"We chose to put a microswitch on the end of the travel, so you had to get your knees out of the way just to do anything at all. You had to move the stick all the way to a stop until it closed the microswitch, opened up a valve, and then the jet worked," Melvill said.

The rudder petals worked the same way, in that they had to be pushed all the way on the floor to activate the thrusters. So, the thrusters would fire as long as the pilot pressed against the microswitches.

Apogee to Atmosphere

Burt Rutan described the idea of the feather maneuver as the pivotal piece of the puzzle needed for the design of *SpaceShipOne*. Reentry into Earth's atmosphere was the most critical point of every spaceflight. By deploying the feather mechanism, referred to as "carefree" and "hands-off," the aerodynamic drag increased substantially, resulting in very low thermal loads. This was because the spaceship slowed down so quickly in the upper atmosphere that when it reached the thick atmosphere, it was traveling with much less energy. Because there was less energy, there was less heat being generated, and *SpaceShipOne* didn't get as hot. On ascent after the boost, as *SpaceShipOne* continued to slow and close in on apogee, the test pilot put the feather up. "We put the feather up because we want to have as much time as possible to troubleshoot if it doesn't

go up," Melvill said. "It goes up with two different pneumatic actuators, either one of which can do the job. They are fed out of two separate high-pressure bottles. And you can put both bottles to one or the other. We had redundancy."

Going up to space had its challenges, but coming back down was where a space program was truly tested. And the feather was *SpaceShipOne*'s ticket back home. Figure 3.15 shows *SpaceShipOne* with its feature in the extended position.

"So, we would put it up as soon as we were out of the atmosphere because if we put it up in the atmosphere, we would start doing loops."

The rear half of each wing folds upward about a hinge line, looking like a jack-knife. It took less than 20 seconds for the feather to deploy to an angle of 65 degrees, and the pilot watched the instrument panel to make sure it went all the way up.

"As soon as you moved the handle, that unlocked it, and as soon as it was unlocked, you'd hear it," Melvill said. "Without the motor running, it was very quiet. And you would hear it go *konk* in the back as it unlocked. As soon as we heard that, you would put the feather up with the handle. There were two handles right next to each other on the left side. The feather took a long time, and it made a noise going up. You could hear the air hissing into the large-diameter actuators."

After *SpaceShipOne* reached apogee, going over the top, it began to pick up speed and continued to follow a ballistic arc downward much

Fig. 3.16. The feather extends and retracts by air-powered actuators, or pistons, that are attached on either side of the fuselage and connected to each side of the wing near their trailing edges. *Mojave Aerospace Ventures LLC, video capture provided courtesy of Discovery Channel and Vulcan Productions, Inc.*

like the parabola traced by a ball thrown into the air and as it drops back down to the ground.

The spacecraft had a lift-to-drag ratio of about 0.7 in the feathered configuration, so the descent was nearly vertical, with an angle of attack of 60 degrees. While in space, a video camera mounted in *SpaceShipOne*'s tail boom captured the image of the feather shown in figure 3.16.

Now as *SpaceShipOne* transitioned from space to reentry, the atmosphere began to get thicker and thicker. Mach 3.25 was the fastest reached by *SpaceShipOne*. This corresponded to an airspeed below 160 knots equivalent airspeed.

"You come back a little faster than you go up," Melvill said. "You get a tremendous amount of g-forces on your body when you are coming back. We were looking at 5.5 g's on reentry."

The pilot didn't wear a g-suit. But he did experience a deceleration force above 5 g's for around 10 seconds. Since the pilot sat upright the entire flight and *SpaceShipOne* reentered the atmosphere belly first, it was critical for the pilot to train beforehand in order to build up his g-tolerance.

"You see because of the rods and cones in your eyes," Melvill explained. "They need oxygen-enriched blood to feed them. If the blood gets drained down to where there is not much oxygen-enriched blood behind your eyes, you go blind. You just black out. And you can still hear and think and move the stick. But you can't see. And that would be hard to fly if you couldn't see."

In the feather configuration, *SpaceShipOne* acted like the conical-shaped shuttlecock, or birdie, used in the sport of badminton. Originally made of feathers and now more commonly made of plastic, the shuttlecock's skirt has such high drag compared to its base that after a hit from a racket, the shuttlecock automatically orients itself base-first in the direction of flight. The Scaled Composites team took advantage of the two important aspects of this concept when considering the reentry of *SpaceShipOne*. The high drag caused rapid deceleration, and the self-aligning tendency ensured the proper orientation.

"You don't even need to do anything coming back down. It is a 'carefree' reentry," Melvill said. "You could put your hands behind your head and take your feet off the rudder pedals and just wait. And you could reenter in any attitude. You could be tumbling when you reenter. You could be upside-down when you reenter. You could be knife-edge and the feather will turn you around and straighten you out. It happens real slowly."

A terminal velocity of 60 knots equivalent airspeed is reached in this high-drag configuration. This corresponds to a ballistic coefficient, calculated by using the weight, drag, and cross-section of *SpaceShipOne*, of 12 pounds per square foot (psf), compared to that of 60 psf for the early Mercury capsules. A low value of the ballistic coefficient means that the spacecraft will begin to slow down quickly in the thin atmosphere. So, *SpaceShipOne* experiences only low overall structural and thermal loading. It goes from supersonic to subsonic in about a minute and a half.

"The temperature when we reenter around the airplane is very high," Melvill said. "It is about 1,200 degrees, but that's the air temperature against the skin. Because that happens at 100,000 feet [30,480 meters] or more, there is so little air to conduct the heat into the structure. The molecules of air are so far apart because it is only 1 percent, or less, atmosphere up there. So, it takes time for that heat to be conducted into the structure, and we're through that heating period before it has time to get into our airplane. Burt designed it that way, and that was very clever. He made sure that we wouldn't spend very much time under the conditions where we could melt the airplane."

During a presentation at the Experimental Aircraft Association's 2006 AirVenture at Oshkosh, Wisconsin, Rutan described the difference between the thermal protection system of *SpaceShipOne* and the Space Shuttle: "You don't have the problems going Mach 4 that you do going Mach 25.

"On the boost, *SpaceShipOne* sees temperatures that are too hot for the skin at the nose and on the leading edges. That's all. To be conservative, we protected some of the areas that got relatively hot on reentry. That's why you see that stuff down under the nose and up underneath the wing. However, our measurements there showed that none of that was required. *SpaceShipOne* doesn't need any thermal protection at all for reentry. It only needs several pounds of material for boost."

In cooperation with the U.S. Air Force, *SpaceShipOne* reentered into restricted airspace controlled by Edwards Air Force Base. But within the restricted airspace, the Office of Commercial Space Transportation (AST) had designated a location, a box roughly 2.5 square miles (6.5 square kilometers) in size, for *SpaceShipOne* to come down through during reentry.

This was certainly a piloting challenge because in order to reenter through the box, the pilot already had to be in position on the way up. So not only did the pilot have to make sure the wings were level, the nose was pointed up, asymmetries were compensated for, and the occasional wind shear was counteracted, he had to try to position *SpaceShipOne* so that once the engine shut down and the atmosphere was gone, it would coast more than 100,000 feet (30,480 meters)

Fig. 3.17. *SpaceShipOne* returns from space as a glider and makes a horizontal landing similar to way the Space Shuttle does it. Although not the most efficient glider, *SpaceShipOne* had a glide range of around 60 miles (97 kilometers). *Mojave Aerospace Ventures LLC, photograph by Scaled Composites*

feet to apogee, free fall back down, and then drop through a relatively small-sized box.

"Our priorities were we wanted to get altitude, and we wanted to leave the atmosphere without a lot of body rates or gyrations," Binnie said.

The third goal was to come back inside the box. "But controlling the body rates and maneuvering the vehicle to find that box were kind of at odds with each other."

SpaceShipOne continued to descend with its feather up. This configuration was so stable that in the atmosphere at the higher altitudes, it was easier for the pilot to just leave the feather up, even though *SpaceShipOne* had performed a safe reentry. However, there wasn't much control. The pilot couldn't pitch the nose and roll the wings, but he was able to change the direction that the nose pointed.

"This is something we didn't feel necessary to test, but it is likely that you could survive a feather-up landing in *SpaceShipOne*," Rutan said. "We did not plan to ride it down if the feather didn't come down. We planned to jump out."

At an altitude below 70,000 feet (21,340 meters), the feather was retracted and locked. *SpaceShipOne*, flying subsonically, transformed into a glider.

Gliding to Mojave

With the wings returned to the normal flight configuration, *SpaceShipOne* became a glider. The hard part was certainly over, and the pilot had time to take a breath and take in the view again. But his work was not completely over. Figure 3.17 shows *SpaceShipOne* gliding over the high desert of Mojave.

If *SpaceShipOne* was off course during the boost phase, it could be far away from where it needed to land. However, the spacecraft had a glide ratio of seven to one. So, *SpaceShipOne* had glide range of about 60 miles (97 kilometers) after it defeathered. "It's got an awful lot of capability to deal with poor trajectory," Doug Shane said.

The pilot also had to resolve a technical glitch with the global positioning system (GPS) receiver. It would drop out or lose its way during spaceflights. "The GPS receiver was never previously tested in that high and in that fast of a flight regime," Pete Siebold said. "And so it had software difficulties of its own. The GPS receiver was something you buy from a company off the shelf. It just didn't perform the way it was supposed to."

In one of the spaceflights, the GPS receiver reset by itself, but for the other two, the pilot had to reset it.

"So, we had to do a power cycle. The avionics go away while it is booting back up, and then it does a realignment of the inertial navigation system once it powers up again. But we could live with that fault. We had workarounds," Siebold said.

The spaceship glided down for 10–15 minutes and was much lighter now that the oxidizer and fuel were burned off. *SpaceShipOne* was not able to land safely with a full load of oxygen and fuel. The extra weight changed the balance, and it was just too heavy for the landing gear to take. So, for an abort, it would have to dump all the nitrous oxide, but it still had to manage with the remaining mass of rubber.

Although *SpaceShipOne* did a good job gliding down and getting close to the airport, it did not have all the controls or responsiveness of a typical glider, so its maneuvering when it came to landing was limited. Early in the program, a few landing attempts were almost too short or too long for the runway.

Pete Siebold and Brian Binnie modified the landing technique, allowing *SpaceShipOne* to easily compensate for coming in too high or too low. "We would fly at 8,500 feet [2,590 meters] above sea level above our touchdown point," Siebold said. "And we had a 360-degree turn to make back to that point again, and then we would be lined up for the final touchdown on the runway. The original technique allowed you to vary the radius of that turn. If you were too low, you could decrease the radius, and your circumference was your flight path. And if you were too low, you could make up for being low on energy by flying that tight radius. Or you could widen it out.

"We also had one last-ditch effort to make any adjustments, and that was to put the landing gear down. When the landing gear was up,

Fig. 3.18. Landing proved to be a bigger challenge than anyone had anticipated. There was only one shot at it. *SpaceShipOne* had to come in at the right altitude and speed or it risked overshooting or undershooting the runway. *Mojave Aerospace Ventures LLC, video capture provided courtesy of Discovery Channel and Vulcan Productions, Inc.*

it was a seven-to-one glide ratio. With the landing gear down, it was a four-to-one glide ratio. The problem was that once you put it down, you couldn't put it back up. So, you had to be sure that you had sufficient elevation to make the runway."

SpaceShipOne would spiral in for a landing while reaching key altitude points that were provided by the TONU. An energy predictor similar to what was used during boost showed the pilot where *SpaceShipOne* would be at the key altitudes based on the current turn and descent rates. The pilot would then adjust his speed and turn so that *SpaceShipOne* would end up at the place it needed to be.

"After we developed that and utilized it, we landed to within 500 feet [150 meters] of a given touchdown point on every subsequent flight. That was real rewarding," Siebold said.

SpaceShipOne approached the runway at an airspeed of 140 knots indicated airspeed. But in order to put its gear down, it had to perform a special maneuver. "There were other peculiarities with the gear system," Siebold explained. "You couldn't put it out at your normal approach speed. So, the speed at which you flew the pattern was too fast to put the gear out and too fast to land. So, what you had to do was in your turn from base to final, you actually had to pull the nose up, slow the airplane down, put the gear out, dump the nose, with gear extension at 125 knots, and then speed back up to 140 knots."

Figure 3.18 shows *SpaceShipOne* gliding down to the runway at Mojave Airport.

The pilot had one last challenge to face. As it turned out, it was one that Charles Lindbergh faced seventy-seven years previously in

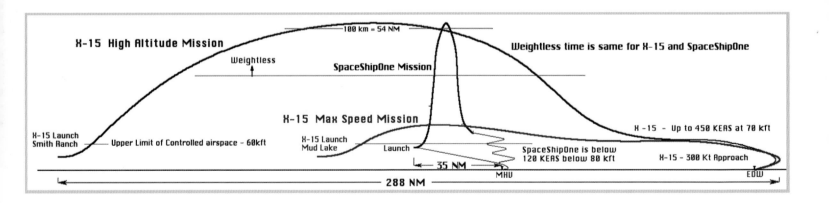

Fig. 3.20. The X-15 had enough fuel to power its rocket engine for about 2 minutes, so it required a B-52 to lift it to launch altitude. The X-15 flew from 1959 to 1968, posting a top speed of Mach 6.70, or 4,520 miles per hour (7,270 kilometers per hour), and a maximum altitude of 354,200 feet (108,000 meters) on separate flights. *NASA–Dryden Flight Research Center*

Table 3.2 *SpaceShipOne's* and X-15 Suborbital Mission Comparison

	SpaceShipOne	X-15
Program goals	Altitude and view	High speed and altitude
Number of vehicles in program	1	3
Crew capacity	3	1
Number of rocket-powered flights	6	199*
Combined time of rocket-powered flights (hours:minutes:seconds)	2:07:26	30:13:49*
Number of flights above 100 kilometers (62.1 miles/328,000 feet)	3	2
1st stage	*White Knight* carrier aircraft	NASA B-52 carrier aircraft
Separation altitude	47,000 feet (14,330 meters)	45,000 feet (13,720 meters)
Engine type	Hybrid	Liquid
Oxidizer and fuel	Nitrous oxide and rubber	Liquid oxygen and anhydrous ammonia
Maximum engine burn time	84 seconds	141 seconds (high-speed mission)
Trajectory for boost and reentry	Nearly vertical	Approximately 40 degrees (high-altitude mission)
Maximum airspeed	Mach 3.25 (on reentry)	Mach 6.70
Maximum altitude	367,500 feet (112,000 meters)	354,200 feet (108,000 meters)
Weightless time	3.5 minutes	3.5 minutes
Reentry method	"Carefree" (60-degree angle of attack)	Pilot controlled pull-up
Reentry max q	80 psf (160 KEAS)	1,000 psf (550 KEAS)
Approach airspeed	140 KIAS**	270 KEAS
Touchdown airspeed	105–110 KIAS	180 KEAS
Landing surface	Runway	Lake bed
Number of vehicles lost during flight testing	0	1
Number of fatalities during flight testing	0	1

*In two of these flights, the rocket-engines were not ignited. One was a planned glide flight while the rocket-engines failed to ignite for the other.
**Knots indicated airspeed

the *Spirit of St. Louis*, which had no front windshield. "The visibility out of *SpaceShipOne* is pretty restricted, and you got these really small windows, and there is no window in front," Melvill said. "So, when you are lined up with the runway, you can't see the runway. With a normal airplane, you can look out the front and see the runway.

"This one, the windows were on the sides, and as long as you were turning toward the runway, you could see it through the side window. But as soon as you lined up with the centerline, you couldn't see it anymore. The whole airport disappeared. So, that was a little bit disconcerting I think for all of us. That's why we had a chase plane sitting right on the wing calling out how high we were above the ground and basically keeping us straight as well."

At 100 to 110 knots equivalent airspeed, the main landing gear at the rear hit first, and then the nose skid followed. There was no real way to steer once it touched down. The wooden tip of the nose skid brought it to a smooth, but slightly smoky, stop in front of an ecstatic crowd. Figure 3.19 shows *SpaceShipOne* being towed from the runway accompanied by Paul Allen, Burt Rutan, and Sir Richard Branson.

Between Spaceflights

Preparing *SpaceShipOne* for its next spaceflight was a relatively simple task that required only minimal maintenance. The spent components of the rocket engine were replaced with fully fueled components, and the oxidizer tank was refilled. The air bottles used to activate the feather and run the reaction-control system and other systems had to be recharged. The ablative coating for the thermal protection system was restored. And since every flight was an envelope expansion, the TONU was updated after a thorough review of the flight data.

X-15 Comparison

The North American X-15 was the first of three winged vehicles ever to have reached space, the Space Shuttle and *SpaceShipOne* being the other two. The basic mission profile was similar for these vehicles in that they all used two stages to reach space and glided back to Earth for an unpowered landing. However, the X-15 and *SpaceShipOne* shared much more in common compared to the Space Shuttle, which was a fully operational spacecraft designed to transport large payloads back and forth from orbit, whereas the other two were research and proof-of-concept vehicles that only reached suborbital altitudes.

Originally conceived in 1954, the X-15 first flew in 1959 and flew the last time in 1968. Its two primary goals were to fly at Mach 6—hypersonic speeds begin at Mach 5—and reach an altitude of 250,000 feet (76,200 meters). The X-15's 199 powered flights directly influenced the Mercury, Gemini, Apollo, and Space Shuttle space programs as well as the U-2 and SR-71 reconnaissance aircraft.

Table 3.2 shows a comparison between the X-15 and *SpaceShipOne*.

To conserve fuel, the X-15 was dropped from the wing of a NASA B-52 carrier aircraft at an altitude of 45,000 feet (13,720 meters), as shown in figure 3.20. For its high-altitude mission, the rocket engine burned for up to 2 minutes, and then the X-15 returned from space and glided in for a landing.

This trajectory flown by the X-15 was, however, quite a bit different from that flown by *SpaceShipOne*. Figure 3.21 shows the trajectory flown by *SpaceShipOne* and compares it with both the high-speed and high-altitude trajectories of the X-15.

The most apparent difference was in the profile width of the X-15 high-altitude mission and the *SpaceShipOne* mission. The X-15 had to cover 331 miles (533 kilometers) in order to reach an apogee above 62.1 miles (100 kilometers), whereas *SpaceShipOne* only needed 40 miles (64 kilometers) to accomplish the same feat.

Both had the same amount of weightless time and view, but the X-15 traveled much faster to achieve this. The higher speeds meant greater aerodynamic loads and thermal protection requirements. But the most important difference was that the X-15 had to expend much more energy to reach the same altitude. The greater the energy needed to get from point A to point B, the more expensive it is to fly.

Primarily constructed of lightweight, high-strength titanium, it had skin of Inconel X, a chrome-nickel alloy that would withstand temperature as high as 1,200 degrees Fahrenheit. The black coating helped dissipate heat, and it was necessary to design gaps into the fuselage to allow for temperature expansion, which was a feature carried over to the SR-71.

A liquid oxygen oxidizer and an anhydrous ammonia fuel powered the liquid rocket engines, providing a thrust of 28,000–57,000 pounds-force (125,000–254,000 newtons). A reaction-control system that used hydrogen peroxide thrusters on the nose and wings allowed the X-15 to maneuver outside the atmosphere.

Another similarity was the landing gear. To reduce weight and simplify the design, the X-15 used two landing skids at the rear of the vehicle, compared to the single skid at the nose used by *SpaceShipOne*.

On August 22, 1963, the X-15 set an altitude record of 354,200 feet (108,000 meters). Four years later, on October 3, 1967, a highly modified version renamed the X-15A-2 set a speed record of Mach 6.70, or 4,520 miles per hour (7,270 kilometers per hour). The entire aircraft had to be covered in a white ablative coating to increase the thermal protection of the skin up to 2,000 degrees Fahrenheit. This was a peak speed, and the X-15 could only run its engine for about two minutes. However, if this speed could be maintained, it would be possible to travel the distance from New York City to Los Angeles in just over a half an hour.

Although it was a tremendously successful program, four major accidents occurred. One of them claimed the life of test pilot Michael Adams due to a control-system failure during reentry. This accident and the Space Shuttle *Columbia* accident, also occurring during reentry, were key influences that drove Rutan to develop the "carefree" feather reentry.

Although *White Knight* began flying about a year before *SpaceShipOne*, construction of both vehicles began at about the same time. High-strength, lightweight composites of carbon fiber/epoxy were used to build the primary structure of both vehicles. *Tyson V. Rininger*

Chapter 4

SpaceShipOne Construction

At the front of *SpaceShipOne's* stout fuselage, many small portholes take the place of a conventional canopy or windshield, giving the vehicle its truly far-out look. As shown in the views in figure 4.1 and figure 4.2, the twin tail booms are another of the spaceship's most distinguishing features. In terms of function, however, *SpaceShipOne's* feather mechanism is unique among all aircraft and spacecraft. The forward half of each wing is fixed, but the rear halves, including the tail booms, fold upward for reentry.

The appearance of *SpaceShipOne* bears resemblance to aspects of several pioneering rocketcraft. The bullet-shaped fuselage appears very similar in shape to the Bell X-1, shown in figure 4.3. However, the X-1 itself shares a common shape with the V-2 rocket, which was initially modeled after a rifle bullet. The use of a delta wing and stabilizers at the wingtips is also reminiscent of NASA's early lifting bodies, as shown in figure 4.4.

Burt Rutan's innovative use of composites took shape in the 1970s when he built his second aircraft, the VariEze. Now with *SpaceShipOne*, an aircraft so radically different in function, purpose, and performance, Rutan and the Scaled Composites team had to tap deep into their expertise. And when this wasn't sufficient, they had to risk taking a leap. Their decades of experience with the manufacturing of strong, lightweight composite aircraft would be tested to the limits, because so much of the design and construction was brand-new territory. So how does one go about building a spaceship? This is a question that wouldn't have

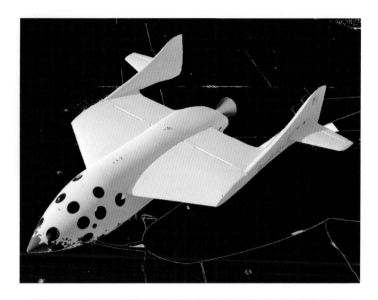

Fig. 4.1. Among *SpaceShipOne*'s most distinct features are the round windows on its bullet-shaped nose, the outboard tail booms at the wingtips, and the thick, swept-back wings mounted high on the fuselage. *Mojave Aerospace Ventures LLC, photograph by Scaled Composites*

Fig. 4.2. At a length of 28 feet (8.5 meters), *SpaceShipOne* is shorter than the Bell X-1, the very first X-plane. However, the width of *SpaceShipOne*, the distance between the tips of the horizontal stabilizers on the tail booms, is 27 feet (8.2 meters), which is comparable in size to the wingspan of the X-1. *Mojave Aerospace Ventures LLC, photograph by Scaled Composites*

Fig. 4.3. Parked in front of the B-29 mothership, the X-1 was built by Bell Aircraft Corporation for the U.S. Army Air Forces and the National Advisory Committee for Aeronautics, the predecessors of the U.S. Air Force and NASA, respectively. The X-1's revolutionary use of structures and pioneering aerodynamic shapes and controls enabled Chuck Yeager to become the first to break the sound barrier, flying faster than Mach 1. *NASA–Dryden Flight Research Center*

Table 4.1 Size Comparisons for Rocketcraft

	SpaceShipOne[a]	X-1	X-15[b]	Space Shuttle
Length	28 feet (8.5 meters)	30.9 feet (9.4 meters)	51 feet (15.5 meters)	122.2 feet (37.2 meters)
Wingspan	16.4 feet (5.0 meters)[c]	28 feet (8.5 meters)	22 feet (6.7 meters)	78.1 feet (23.8 meters)
Height	8.8 feet (2.7 meters)	10.8 feet (3.3 meters)	13 feet (4.0 meters)	56.6 feet (17.3 meters)
Weight[d]	7,937 pounds (3,600 kilograms)	12,250 pounds (5,557 kilograms)	38,000 pounds (17,237 kilograms)	242,000 pounds (110,000 kilograms)

a: For last spaceflight of *SpaceShipOne*.
b: For modified X-15A-2 without drop tanks.
c: *SpaceShipOne*'s width of 27 feet (8.2 meters) is its widest dimension.
d: Gross weights are given except for the Space Shuttle, which is given for landing weight.

an answer until *SpaceShipOne* was ready for flight testing. Even then, after each step forward and envelope expansion, it was necessary to make modifications or refinements to overcome the technical challenges that waited in the wings.

Overall Dimensions

At 28 feet (8.5 meters) in length, 8.8 feet (2.7 meters) in height, and 27 feet (8.2 meters) in width, which is the distance between the tips of *SpaceShipOne*'s horizontal stabilizers, the spacecraft is only slightly smaller than the Bell X-1. Table 4.1 shows a size comparison between *SpaceShipOne*, the X-1, the X-15, and the Space Shuttle. Although the width of *SpaceShipOne* is slightly wider than the wingspan of the North American X-15, it is about half the length of the X-15. Every time *SpaceShipOne* flew, it had a different weight. The final spaceflight was

the heaviest, with an empty weight of 2,646 pounds (1,200 kilograms) and maximum weight of 7,937 pounds (3,600 kilograms). The weight advantage of *SpaceShipOne* was clear when compared to the gross weights of 12,250 pounds (5,557 kilograms) and 38,000 pounds (17,237 kilograms) for the X-1 and X-15, respectively.

Fuselage and Composite Structures

The construction of *SpaceShipOne* really began with the building of the fuselage. Without its wings and tail attached to the fuselage, *SpaceShipOne* looks like a stubby little rocket. Figure 4.5 shows the crew compartment in the forward section of the fuselage, the oxidizer tank in the middle section, the rounded pressure bulkhead that separates the crew compartment from the oxidizer tank, and the rocket engine in the aft section. The maximum outer diameter of the cylindrical

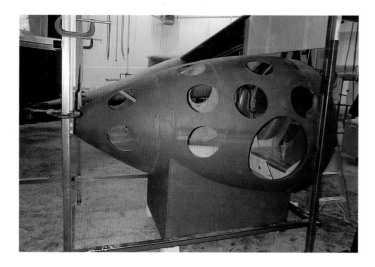

Fig. 4.6. A first step in the assembly was to build the carbon fiber/epoxy composite subassemblies that made up the fuselage. The subassemblies were bonded together, except for the nose cone, which was a detachable emergency escape hatch. *Mojave Aerospace Ventures LLC, photograph by Scaled Composites*

fuselage is 60 inches (152 centimeters). It is a monocoque design, which means that the fuselage hull provides most of the structural support and load bearing for the spacecraft. However, the rocket engine obviously exerts a tremendous amount of force. So, the oxidizer tank is actually a very important structural member as well.

SpaceShipOne was put together much the same way a small plastic model airplane is put together. Beginning with a bunch of individual parts, they are assembled piece by piece. Woven layers, also called plies, of carbon fiber and epoxy were the primary materials used to make most of the lightweight composite parts of *SpaceShipOne*. These parts were used to build the fuselage, the wings, and the tail booms.

The whole process began by designing the parts using three-dimensional computer-aided design (CAD). The designs were then fed into an automated computer numerical control (CNC) machine to carefully shape the parts by whittling down foam blocks. Even though these foam parts precisely resembled the actual composite parts, they did not share the same strength, durability, resilience, and imperviousness to temperature. The foam parts were used to create molds that would then be used to create the actual composite parts to build *SpaceShipOne*.

After the molds were created, lay-up began. Here the composite material was built up layer by layer. Once enough material was added and the layers were to the proper thickness, the parts were vacuum-bagged and oven-cured. The vacuum-bagging process is basically what it sounds like. A part is covered with an airtight bag, which is evacuated by a pump. What this does is remove air from in between the layers and volatile compounds that are in the epoxy, which have become unwanted byproducts. The more thoroughly they are removed, the stronger the composite will be. The oven-curing enables the epoxy to properly set and achieve the desired properties. After the parts for *SpaceShipOne* were cured, assembly began.

Although the materials may have changed, the building technique was not unlike that used to build European sailplanes over the past forty or fifty years.

The primary structure of the fuselage is made up of only a few very large subassemblies. Figure 4.6 shows the nose cone and cabin section. The subassemblies have edges that fit to one another. They are then fastened with a jig and chemically bonded together. The subassemblies

Fig. 4.7. Piece by piece, the exterior of *SpaceShipOne* took shape, like the building of a plastic model airplane. As construction proceeded, the wings were attached to the fuselage, and all the wiring, plumbing, linkages, and other components were installed inside. *Mojave Aerospace Ventures LLC, photograph by David M. Moore*

Fig. 4.8. Because the cockpit was pressurized and *SpaceShipOne* faced extreme conditions in space, the walls of the fuselage provided double containment. In between the "shell within a shell" was an insulating layer that also improved the structural strength of the fuselage. *Mojave Aerospace Ventures LLC, photograph by David M. Moore*

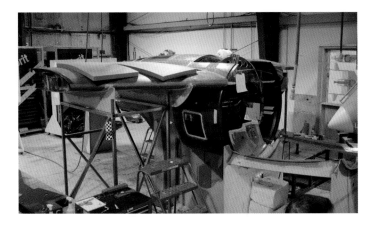

Fig. 4.9. This photograph shows the early stages of construction for the aft section of *SpaceShipOne*. A pressure bulkhead will be put in place at about the location of the opening in the fuselage to separate the oxidizer tank from the cockpit. *Mojave Aerospace Ventures LLC, photograph by David M. Moore*

together form the inner hull or shell of *SpaceShipOne*. This inner shell is still not strong enough to support the spacecraft.

The nose cone is attached to the rest of the fuselage a little differently. The nose cone is the primary escape hatch. Its edge is keyed so that it can lock and unlock from the fuselage with a quick little turn.

Figure 4.7 shows the cabin section that is forward of the pressure bulkhead in various stages of construction. Figure 4.8 shows the top of the cabin section after even more assembly, and figure 4.9 shows the aft section where the wings mount.

The next step in building the fuselage was to add a core of honeycomb material on top of the inner shell. Nomex, made by DuPont, was used as the honeycomb core. Not only is the honeycomb core lightweight, it offers high strength and it is heat and fire resistant. The honeycomb core, however, cannot serve as the outer layer. It needs further covering. So, panels called skins, manufactured using this composites process, were attached to the honeycomb core. The fuselage is a shell within a shell. The process of adding successive layers creates what is known as a sandwich structure, as the fuselage hull is viewed in cross-section. Figure 4.10 shows an example of a sandwich structure. Another way to look at this is that the fuselage hull is a thick composite made up of several thin composites.

Honeycomb Sandwich Structure

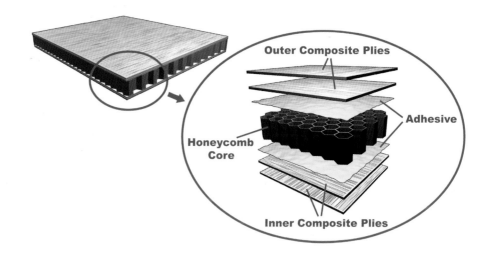

Fig. 4.10. This diagram shows a honeycomb sandwich structure that is representative of some of Burt Rutan's designs, including *SpaceShipOne*. Here, for example, composite plies sandwich a core material shaped like honeycombs in order to provide extremely good strength-to-weight properties. *James Linehan*

After the assembly of the composite parts, *SpaceShipOne* resembled a spacecraft, but the job was far from over. "The hard part is stuffing all the systems in it and making sure that they are mounted properly," Rutan said.

Initially the fuselage ended at the throat of the rocket-engine nozzle. During the flight tests, a fairing was added to extend the fuselage to the rim of the nozzle in order to improve aerodynamics.

Crew Compartment

SpaceShipOne has a space-qualified environmental control system (ECS). Its pressurized cabin has room to fit three people. The pilot sits up front in the nose, and behind him and up against the pressure bulkhead is a row of two seats for the passengers.

The pilot and the passengers sit upright but slightly reclined, as shown previously in figure 4.5. This helps them tolerate the g-forces they face during the boost and reentry phases of the mission. They do not have to wear spacesuits or g-suits, but *SpaceShipOne* has an oxygen system with oxygen masks for them to wear.

The backseat row is less than 2 feet from the oxidizer tank, but the pressure bulkhead separates the cabin from the oxidizer tank and the rest of the rocket engine. The dome shape of the pressure bulkhead is necessary. "These shapes are real important as pressure vessels," Rutan said. "And it is pressurized all the way to the nose. There is not another bulkhead up in front."

For test flights, the pilot pressurized the cabin to 4,000–6,000 feet (1,220–1,830 meters). An airliner sets the pressure inside its cabin to about 8,000 feet (2,440 meters). This means that no matter how high or low it flies, the passengers inside will always feel a pressure equal to what they would feel if they were standing on a mountain with a height above sea level of 6,000 feet (1,830 meters).

The cabin was sealed but did have a small amount of leakage. The pilot watched the cabin altimeter, which was used to measure the cabin pressure, and manually adjusted it as necessary.

SpaceShipOne does not have its own heating or cooling system. During captive carry, however, the vehicle was heated by bleed air from *White Knight*'s engines, which pumped the hot air to the pressure bulkhead. The cabin temperature did not change by more than 15 to 20 degrees Fahrenheit from the time the door was closed on the ground. Again, the short duration of the mission really played to Rutan's design principles of simplicity and low construction cost.

At low altitudes, the pilot could get fresh air by opening two 4-inch (10-centimeter) plugs located on either side of the fuselage. Similar to the design of airliner doors, the plugs open inward and are beveled, like corks, so that the high pressure inside the cabin helps keep them closed tight and prevents opening at high altitudes. "You didn't need cooling," Doug Shane said. "You could keep the airplane cool on the ground with the [external] air conditioner. Once you took off, you could wait to put the plugs in until you were at 10,000 to 12,000 feet [3,050 to 3,660 meters], where it is pretty cool."

The plug on the pilot's right-hand side has a safety pressure relief valve that could be capped in case it failed. The other plug has a manual ball valve that opens to dump the pressure in the cabin if necessary. This plug also has a big tab riveted to it. In an emergency situation where the crew would have to bail out, they would have to wait for the cabin to depressurize through a small valve. The tab provided the leverage so the pilot could peel off the plug fast, allowing the cabin to rapidly depressurize. Once *SpaceShipOne* was all sealed up, it was essentially a trapped volume of air. "There's no exchange of air. So you've got to be concerned about humidity and carbon dioxide," Shane said.

A second hose coming off each oxygen mask collected the exhaled air in order to control the carbon dioxide (CO_2) levels and humidity. The exhaled air was dried, and a scrubber using an absorbent material was used to remove excess CO_2. But because of the mission's short duration and the fact that the cabin was sealed off from the atmosphere

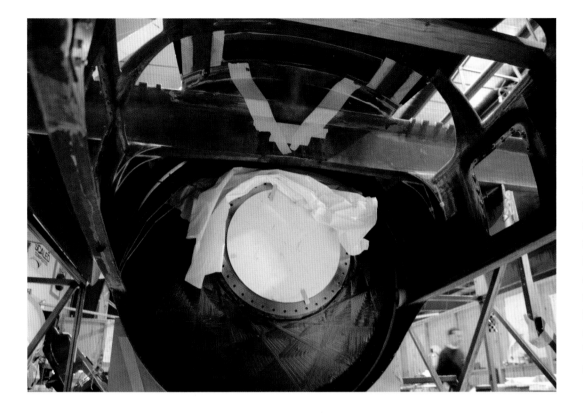

Fig. 4.11. Looking from underneath into the engine bay, these are the two main spars running perpendicular through the fuselage, one in front of the oxidizer tank and one behind it. These provided the structure to support the fixed and movable sections of the wings. *Mojave Aerospace Ventures LLC, photograph by David M. Moore*

Wings

The design of the wings had to take into consideration more factors than most other winged aircraft must consider. The wings of *SpaceShipOne* had to perform from subsonic to supersonic, withstand reentry into Earth's atmosphere, and incorporate the mechanism of movable wings. No other winged vehicle has had to tackle all these at the same time.

Swept wings, which look like delta wings with the wingtips cut off, are attached high on the fuselage. This shape was required for supersonic flight. A tail boom with an outboard horizontal stabilizer is mounted to each wingtip.

The wings have an airfoil shape, but a hinge runs along the full length of the wingspan. The hinge allows the aft section of the wings to pivot up for the feather maneuver and back down after reentry. The forward wing sections, which are roughly the front two-thirds of each wing, do not move.

The wing area is approximately 160 square feet (15 square meters) and the wingspan is 16.4 feet (5.0 meters). However, since the horizontal stabilizers of the tail booms extend out farther than the wings, the width is 27 feet (8.2 meters).

The aspect ratio of the wing is 1.7, which is very low compared to the high aspect ratios of the long, thin wings of sailplanes. For traditional gliders, the lift characteristics are a key design factor. But for *SpaceShipOne*, it requires only enough lift to be able to point its nose upward during rocket-powered ascent in the atmosphere and to be able to glide back after reentry to the landing site for a safe touchdown.

There are two main spars that each run through the wing from tip to tip. One spar goes in front of the oxidizer tank, and one goes behind it, which can be seen in figure 4.11.

Fig. 4.12. *SpaceShipOne*'s wings had to be very rigid and strong because they not only supported the tail booms and feather mechanism, they also had to withstand the very high forces encountered during boost and reentry. This photograph shows how thick the ribs inside had to be. *Mojave Aerospace Ventures LLC, photograph by David M. Moore*

for a relatively short time, little CO_2 buildup occurred, and makeup oxygen (O_2) was not necessary.

"We actually demonstrated on *White Knight* that we had adequate control with three people on board for a three- or four-hour flight," Shane said. "We knew it would be fine for one hour."

Fig. 4.13. NASA had designed the Ames-Dryden-1 (AD-1) to explore the use of an oblique wing that could pivot during flight. During takeoff, the AD-1's wing was perpendicular to the fuselage, like a traditional aircraft. In order to evaluate fuel efficiency, it was possible to pivot the wing to a maximum of 60 degrees, as shown in this photograph. NASA contracted Burt Rutan's RAF to analyze design and loading characteristics. *NASA–Dryden Flight Research Center*

"The wing is not tapered in total thickness," Rutan said. "It is as thick at the tip as it is at the root. It has to do with meeting the stiffness to support the boom. And of course for the hinge line of the boom, it has to be a perfectly straight line or it would bind."

There are no control surfaces on the leading or trailing edges of the wings like other aircraft. Wings on most aircraft also store fuel. But since fuel is stored in the rocket engine itself, which runs through the fuselage, and the oxidizer is in a large tank behind the cockpit, the wings have plenty of room to fit other components and systems, as shown in figure 4.12.

Tail Booms

All the flight control surfaces are on the tail booms, which are mounted to the wingtips and pivot with the aft wing sections when the feather is deployed. Each tail boom has a vertical stabilizer and horizontal stabilizer.

Upper and lower rudders are mounted at the back of each vertical stabilizer for yaw control. Pitch and roll is controlled by elevons that are attached to the trailing edges of the outward extending horizontal stabilizers. The fiberglass construction of the elevon skin gives radio transparency for antennas. For control during supersonic flight, the entire horizontal stabilizer on each tail boom pivots.

Minor modifications were made to the tail booms during flight testing. To resolve an aerodynamic problem, the distance from tip to root of the horizontal stabilizers was increased by 16 inches (41 centimeters). Also, a triangular strake was added in front of each horizontal stabilizer, and a flow fence was added midspan on each horizontal stabilizer.

Fig. 4.14. The feather was actually a single, unified piece. The right and left rear wings and tail booms were attached to a very sturdy spar. The front edge of the feather fit into a cove at the rear edge of the forward fixed wings. A hinge that ran the length of the wings connected the feather to the fixed wings. *Mojave Aerospace Ventures LLC, photograph by David M. Moore*

Fig. 4.15. Critical to the safe return from space, the feather had only two positions, all up or all down. Redundant pneumatic actuators raised and lowered the feather. While retracted, the feather was held in place by a redundant locking system. However, only the force from the pressurized actuators was needed to keep the feather fully extended. *Mojave Aerospace Ventures LLC, photograph by David M. Moore*

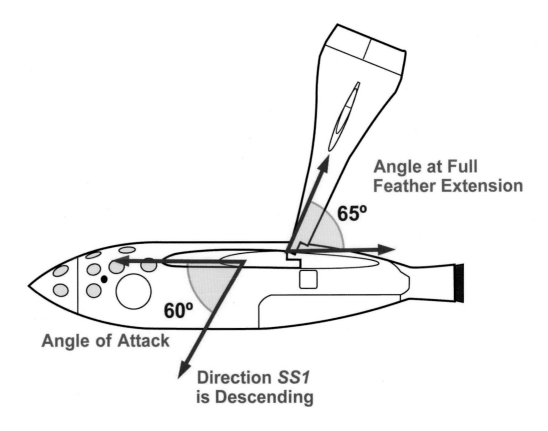

Fig. 4.16. During reentry and with the feather extended at an angle of about 65 degrees, *SpaceShipOne* descended nearly level on its belly. It did not drop straight down, though, but instead moved forward as it fell. The diagram shows the angle of attack at 60 degrees, which is a measure of the direction of motion with reference to the position of the wing. *James Linehan*

The Feather

Mechanically, the most complicated system on *SpaceShipOne* was its feathering system. It was also the most important system on board for ensuring the safety of the pilot and the success of the mission. Rutan already had experience with movable-winged aircraft. His RAF analyzed the designs and loads for NASA's scissor-wing AD-1 (refer to figure 4.13).

Before *SpaceShipOne* reentered the atmosphere, the aft section of both wings, including the tail booms, rose up as if the spacecraft were almost folding in half. With the feather extended, *SpaceShipOne* could reenter the atmosphere with very little pilot input required. This "carefree" reentry was one of the most important elements of Scaled Composites' entire space program.

The feather is a separate structure from the forward wing sections and has its own spars and ribs. However, the aft wing sections and the tail booms do not move independently. Figure 4.14 shows the spar that runs through the fuselage from one end of the movable wing to the other, tying the feather all together.

Left and right pneumatic actuators, which are just cylinders with movable pistons, used air power to pivot the feather up or down along the hinge. The lower ends of each actuator are attached to the fuselage, and the upper ends are attached to the inner face of the aft wing section, as shown in figure 4.15.

There are just two positions of the feather, up or down. The angle the feather makes with the fuselage is preset to 65 degrees, so the pilot did not have to make any adjustments. It took about 13–14 seconds to raise or lower. The feather could be elevated on ascent once the airspeed was less then 10 knots equivalent airspeed. Figure 4.16 shows the 60-degree angle of attack for *SpaceShipOne* descending in the feather condition.

When the feather got to the fully extended position, there was no lock. It was just the force from the actuators that held the feather up. "As it turns out, it takes no load to put it up because you are weightless," Rutan said. "When you are weightless, there are no aerodynamics. It takes no force. In fact, if you would unlock it in space, you could take a little cable and pull it right up. There is no load on it. Okay. But, you need to hold it up for reentry." The reentry force on the flight test actually tried to push the feather down.

If the feather were to move or become disengaged at the wrong time, the results could be disastrous. To keep the feather in the retracted position and to keep it from moving during the other phases of flight, two L-shaped clasps from the locking system secure the trailing edges of the aft wing sections. To unlock the feather so it could be deployed, separate pneumatic actuators were pressurized to retract the clasps.

The design had built-in redundancy for the components that make up the feather system. The two clasps were coupled so they moved together. But they had separate pneumatic sources, lines, regulators, valves, and again, separate actuators. "So, I could have a fire. I could have a line come off. I could have a loss of pressure. I could have all of these things go wrong, and it doesn't affect the other," Rutan said.

The two different pressure sources could run either actuator. The redundancy of the locking system was identical to the redundancy of the elevating systems. They were two independent systems that, under normal conditions, acted in unison. However, either system could engage or disengage the clasps of the other system. Once *SpaceShipOne* was flying subsonically after reentry and the loading drops below 1.2 g, the pilot could retract the feather.

Landing Gear

After a glide test, a rocket-powered flight, or a trip to space, *SpaceShipOne* made a horizontal landing on a runway like most other aircraft. As *SpaceShipOne* got ready to land, the pilot pneumatically actuated the nose skid and rear landing gear, as shown in figure 4.17.

A spring and gravity extended the nose skid into position. It had a maple wood tip that helped slow down the aircraft during landing. This unusual piece of landing gear also acted as a crush damper. Its simple design dramatically reduced the weight and complexity that is typical of retractable nose wheels.

The rear landing gear was also spring and gravity driven but had independent hydraulic brakes for each wheel. By fully depressing a rudder pedal, the brake engaged for the wheel on the corresponding side.

The aircraft was not equipped to retract the landing gear on its own. So, once the pilot put the landing gear down, the only way to get it back up was to land and let the ground crew reset it. There was a big, removable panel on *SpaceShipOne*'s belly where the rear landing gear is located that also provided access for ground support.

Doors and Windows

One of the more distinguishing features of *SpaceShipOne* is its windshield, made of sixteen 9-inch- (23-centimeter-) diameter windows. The windows are small and round to keep the weight low and the structural strength high. Good visibility for the pilot flying *SpaceShipOne*, during all phases of the mission, was an important design consideration. This determined the arrangement of the windows.

With a slight tilt of the head, the pilot could always keep the horizon in sight. For one of *SpaceShipOne*'s rocket-powered flights, this proved crucial when the avionics display went temporarily blank. However, similar to the *Spirit of St. Louis*, the windows do not allow the pilot to see directly ahead of the spacecraft during landing.

Each window has dual panes and dual seals. This redundancy helped prevent loss of cabin pressurization in the case of damage to a window. The outer panes are 5/16-inch- (0.79-centimeter-) thick, heat-resistant Lexan polycarbonate. They are separated by a 1/4-inch (0.64-centimeter) gap from 5/16-inch- (0.79-centimeter-) thick Plexiglas inner panes. There are small vent holes in the outer panes to help prevent the window from fogging up. The inner panes took all the pressurization and when loaded, could deflect 0.2 inches (0.5 centimeters). Even if the inner panes failed, the leak rate would be very low, and *SpaceShipOne* could easily glide back home. Airliner windows also commonly use a two-pane construction with vent holes.

The crew entered *SpaceShipOne* through a 26-inch- (66-centimeter-) diameter, dual-sealed plug door on the port side. The door does not have an external handle but does have an internal handle that the crew could grab and pull out. Just like the plugs on the sides of the cockpit, it is shaped so that the pressure inside the spaceship held the door in place.

The spacecraft was not designed to have ejection seats, in order to help keep the cost, weight, and complexity at a minimum.

The nose cone was an escape hatch. Once it was unlocked, the pilot uses a handle near his left foot to turn the nose cone on its gear ring. After a clockwise turn of only 7.5 degrees, the nose cone detached and fell free from *SpaceShipOne*. Figure 4.18 and figure 4.19 both show views after the nose cone was detached. During an emergency egress, the

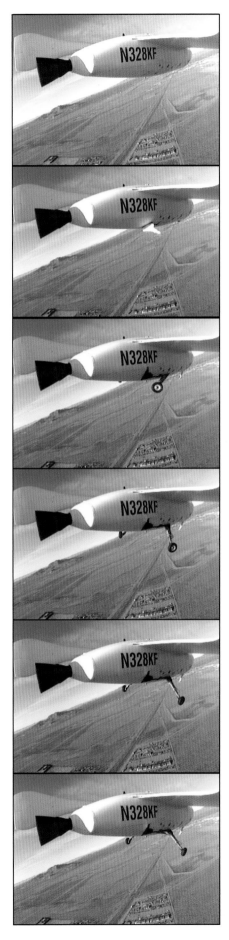

Fig. 4.17. Shot from a camera mounted in the tail boom, these video-capture images show *SpaceShipOne* dropping its landing gear. Both the front nose skid and rear landing gear were spring driven. The only hydraulics used by the vehicle were for the brakes of the rear landing gear. *Mojave Aerospace Ventures LLC, video captures provided courtesy of Discovery Channel and Vulcan Productions, Inc.*

Fig. 4.18. A 36-inch (91-centimeter) opening reveals the cockpit after the nose cone twists off. The crew could use this opening or the 26-inch- (66-centimeter-) diameter plug-style door on the left side of the cockpit for emergency egress if necessary. *Mojave Aerospace Ventures LLC, photograph by David M. Moore*

rudder pedals as well as most of the instruments are dragged out of the cabin by the nose cone, clearing a 36-inch- (91-centimeter-) diameter opening for the crew to crawl through. The crew then would have parachuted to safety after clearing *SpaceShipOne*.

Thermal and Radiation Protection

Because *SpaceShipOne* slowed down so quickly, it did not experience extreme temperatures very long. Therefore, thermal loads were much smaller than those faced by the Space Shuttle. *SpaceShipOne* required only a relatively simple thermal protection system (TPS). Its TPS design consisted of two main parts.

The first part was built in during the manufacture of the composites. When the composite skins for the areas that would experience high temperatures during reentry were constructed, instead of

epoxy, a phenolic resin was used with the carbon fiber. The temperature tolerance for these composites increased by 50 to 70 degrees Fahrenheit.

About 14 pounds (6.4 kilograms) of an approximately 0.035-inch- (0.09-centimeter-) thick ablative coating developed by Scaled Composites was added to 25 percent of the surface of *SpaceShipOne* as the second part of the TPS. Ablative coatings made of reinforced plastic have been around since the early space program. The ablative process reduces the temperature of a spacecraft's surface that faces the airstream on reentry by absorbing some of the heat that is generated. The heat absorbed causes the ablative coating to burn free of the spacecraft, so, in effect, the coating carries away a portion of the heat when it flies off the spacecraft.

When the ablative coating burns, it undergoes a chemical reaction. The heat provides the energy needed for this chemical reaction to

Fig. 4.19. In an emergency, the pilot could remove the nose cone to allow the crew to escape and parachute to safety. A quick turn of a handle near the pilot's left foot detached the nose cone along with the instruments and controls in the way. *Mojave Aerospace Ventures LLC, photograph by David M. Moore*

occur. Therefore, the heat absorbed during the ablation process is heat that is no longer available to heat up *SpaceShipOne*. The ablative coating is then reapplied for the next spaceflight. Figure 4.20 shows the temperature effects on a wingtip and its colored wax test stripes.

Even in the worst-case scenario where the TPS was completely gone, the fuselage could have withstood the damage and returned the crew unharmed. Because of the short duration and relatively low altitude of the spaceflight, *SpaceShipOne* was not equipped with radiation shielding.

Air and Electrical Power

Clean, dry air was used in the pneumatics to pressurize the actuators. The cabin was also pressurized with air. Each system had its own high-pressure bottle and a backup bottle. The feather, environmental control system (ECS), and reaction control system (RCS) each had a bottle A and bottle B. The initial pressure of these six bottles was 6,000 pounds per square inch (psi). Other systems also required pressurized air, but they fed off of these bottles. Electricity was provided by an array of lithium batteries.

SpaceShipOne was the first manned spacecraft to use a hybrid rocket engine, which is a cross between a liquid-fueled rocket engine and a solid-fueled rocket engine. Designed by Scaled Composites, it ran by using a combination of synthetic rubber and nitrous oxide. *Mojave Aerospace Ventures LLC, photograph by David M. Moore*

Chapter 5

SpaceShipOne
Rocket-Engine Design

Spacecraft have used both solid and liquid rockets, and in some cases both, to blast out of the atmosphere, into orbit, to the Moon, and out of the Solar System. The Space Shuttle, for example, uses two solid rocket boosters (SRB) mounted to the external tank (ET) and its three liquid-fueled main engines to reach orbit.

SpaceShipOne had a much different set of challenges to face, so its rocket engine had to be equally unique. There was no off-the-shelf rocket engine that Scaled Composites could simply install. Rutan had to design the rocket engine from scratch. It would be the first that Scaled Composites would have to build. Once the design was complete, Scaled Composites enlisted four subcontractors to provide the rocket-engine components that were not built in-house.

SpaceShipOne would be the first manned spacecraft to use a hybrid rocket engine. Figure 5.1 shows an external view of *SpaceShipOne*'s hybrid rocket engine.

The Rocket Engine

In 1999, Scaled Composites began researching rocket-engine technology. By January of 2000, it had not only identified the type of rocket engine and selected the propellants, but it had developed a new concept for its configuration.

Fig. 5.1. A hybrid rocket engine offers advantages of both liquid-fueled and solid-fueled rocket engines. The rocket engine can be shut off at any time during the burn and can be constructed without complicated plumbing and pumps. The disadvantage, though, is that it has lower performance than the other two types. *Mojave Aerospace Ventures LLC, photograph by David M. Moore*

Rutan believed that the highest risk of the program from the technical stance was the operation of the rocket engine. Reentry was dangerous, of course, but the "carefree" approach using the feather dramatically minimized this danger.

"I ruled out solids because I couldn't do flight tests with them," Rutan said. "I couldn't do flight-test envelope expansion. I couldn't do partial burns. Also, I knew that likely during a burn, I might be accelerating into a Mach number that I'd never been to. And I may not like it. I wanted to be able at any time to shut the motor off just like that.

"I ruled out liquids because they had a large number of failure points that were difficult to improve safely by making them all redundant. If you did, you ended up with a complex system, which historically has been shown to be less safe than not having the redundancy."

A hybrid rocket engine fit Rutan's requirements. It was very safe and very simple and very robust. Just as the name suggests, a hybrid rocket engine is part liquid rocket engine (like the Space Shuttle's main engines) and part solid rocket engine (like the Space Shuttle's solid rocket boosters). Figure 5.2 shows the basic designs of liquid, solid, and hybrid rocket engines.

Essentially, a hybrid rocket engine is a tank that contains the liquid part and a motor that contains the solid part. Upon ignition, the liquid flows into the motor and out come the flames. It can be stopped instantly, unlike a solid, and its propellants are room temperature as opposed to cryogenic. However, there is a tradeoff. Hybrid rocket engines are typically less efficient than liquid or solid rocket engines. This means that for equal amounts of propellant by mass, hybrids deliver less thrust. But in the case of *SpaceShipOne*, the lower performance was acceptable.

"Would I use a hybrid motor to go to orbit? Probably not unless we could develop one that was close to the efficiency of the liquids," Rutan said.

Liquid Rocket Engine

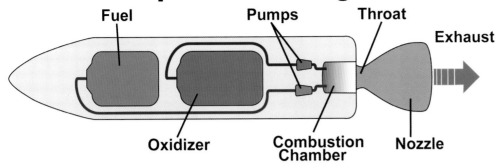

Fuel · Pumps · Throat · Exhaust · Oxidizer · Combustion Chamber · Nozzle

Solid Rocket Engine

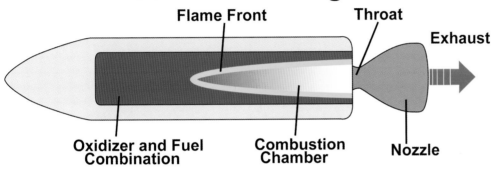

Flame Front · Throat · Exhaust · Oxidizer and Fuel Combination · Combustion Chamber · Nozzle

Hybrid Rocket Engine

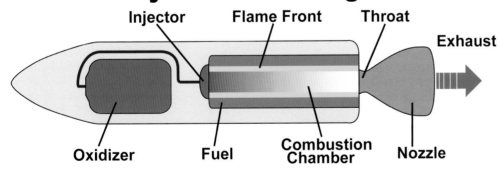

Injector · Flame Front · Throat · Exhaust · Oxidizer · Fuel · Combustion Chamber · Nozzle

Fig. 5.2. The main difference between liquid, solid, and hybrid rocket engines is the state of the fuel and oxidizer used. A liquid rocket engine uses a liquid oxidizer and liquid fuel that are stored separately. The oxidizer and fuel for a solid rocket engine are combined ahead of time to form a solid propellant. A hybrid rocket engine, on the other hand, uses a liquid oxidizer and a solid fuel that mix once it fires off.

James Linehan

Propellants

The force that causes a rocket engine's thrust results from combustion. This type of chemical reaction is similar to burning wood. The wood, which is a fuel, and the air, which is an oxidizer, react to form gases and other substances. Both the fuel and oxidizer must be present for combustion to occur.

The difference with a rocket engine is that the fuel and the oxidizer release much higher energy, and the gases from the reaction travel out of the nozzle at very high speeds. These high-speed gases provide the thrust.

Each Space Shuttle SRB contains 1,100,000 pounds (500,000 kilograms) of solid propellant, 70 percent ammonium perchlorate for its oxidizer and 16 percent aluminum powder for its fuel. The balance consists of binders, curing agents, and catalysts. These burn until there is nothing left to burn. The liquid propellants are contained in separate pressurized tanks within the ET. These tanks hold 1,350,000 pounds (612,000 kilograms) of liquid oxygen oxidizer, often called LOX, and 227,800 pounds (103,000 kilograms) of liquid hydrogen fuel. The main engines produce thrust when the liquid oxygen and liquid hydrogen are pumped together.

Fig. 5.3. To reach orbit, the Space Shuttle uses three liquid-fueled rocket engines, the shuttle main engines (SME) at the rear of the orbiter and two solid rocket boosters (SRB), which are attached to the external tank (ET). *Dan Linehan*

Fig. 5.4. When the oxidizer and the fuel combine, a vigorous chemical reaction takes place. The very hot gases from this combustion produce the thrust and resulting fiery plume. *Mojave Aerospace Ventures LLC, photo provided courtesy of Discovery Channel and Vulcan Productions, Inc.*

Hydroxyl-terminated polybutadiene (HTPB) is the solid used for the fuel and is a synthetic rubber, like that used to make tires. About 600 pounds (270 kilograms) fuel the rocket engine. Both the oxidizer and the fuel can easily and safely be stored and transported. In fact, unlike liquid oxygen and liquid hydrogen, which react together spontaneously, N_2O and HTPB will not react together unless an igniter is first used. For the reaction to occur, the temperature must be greater than 570 degrees Fahrenheit.

The combustion products from the combination of N_2O and HTPB are mostly carbon dioxide, carbon monoxide, hydrogen, nitrogen, and water vapor. This is not as clean burning as the Space Shuttle's main engines but is much less polluting than the Space Shuttle's solid rocket boosters, which produce a giant, toxic acid cloud. Figure 5.3 and figure 5.4 show the rocket engines of the Space Shuttle and *SpaceShipOne* firing, respectively.

Construction Begins

Scaled Composites took what it described as an advanced all-composite design approach to building the rocket engine. "We did develop a new configuration of hybrid rocket motor," Rutan said. "And we patented it. The patent is related to the fact that the whole motor cantilevers off the tank. And doesn't require additional mounting and doesn't build up stresses because of the temperature effects and the aerodynamic loads. It's a key reason that this remains a very simple motor.

"During our development of the new hybrid rocket motor," Rutan added, "Scaled contracted with four companies to provide components." Most of the rocket engine was designed and developed in-house, though.

The two main parts of the rocket engine are the oxidizer tank and the integrated case/throat/nozzle (CTN), as shown in figure 5.5 and figure 5.6, respectively.

The oxidizer tank is a very important structural element of the rocket engine and *SpaceShipOne* as a whole. It is reusable and is bonded to the inside of the fuselage, all the way around, with an elastomeric

SpaceShipOne is a tiny fraction of the Space Shuttle's mass, and it reaches less than a third of the height the Space Shuttles does. So, the propellant requirements are quite different. *SpaceShipOne* used nitrous oxide (N_2O), a colorless liquid or gas naturally occurring in the atmosphere. Nitrous oxide is commonly used as laughing gas, as a hot rod fuel additive for a quick boost of speed, and as a propellant for whipped cream. The N_2O is liquefied and used as the oxidizer. The oxidizer enables the fuel to burn at a near-explosive rate. The oxidizer tank contains 3,000 pounds (1,360 kilograms) of N_2O.

Fig. 5.5. This oxidizer tank stored the liquid nitrous oxide (N_2O). After the tank was filled, it self pressurized, so a pump was not needed to inject the oxidizer into the fuel. The photo shows the inner liner of the oxidizer tank. *Mojave Aerospace Ventures LLC, photograph by David M. Moore*

Fig. 5.6. The CTN (case/throat/nozzle) was a large composite casing filled with synthetic rubber used as fuel. Once the rocket engine was ignited, the exhaust moved through the casing, was compressed by the throat, and then expelled out the nozzle. *Mojave Aerospace Ventures LLC, photograph by Scaled Composites*

Fig. 5.7. The oxidizer tank had a special composite overwrap that allowed it to contain the oxidizer at a high pressure of 750 pounds per square inch (psi). It was permanently bonded to the inside of the fuselage, so it was also an important structural member. *Mojave Aerospace Ventures LLC, photograph by Scaled Composites*

compound. The bond is damage tolerant and isolates vibration, and because of the size of the bond area, the loading due to the force from the rocket engine is distributed over a large area.

The oxidizer tank has a fiberglass liner and two titanium interface flanges, fore and aft. After pressure testing the liner, Scaled Composites contracted Thiokol to provide a graphite/epoxy overwrap using a filament-wrap process, which can be seen in figure 5.7.

The CTN is mounted directly on the rear flange of the oxidizer tank with bolts and an O-ring seal (refer to figure 5.8). This cantilever mounting requires no other additional support, and, by design, reduces the number of potential leak paths.

Fig. 5.8. The CTN attached directly to the oxidizer tank. This photograph shows the mounting ring used to bolt it on. The hybrid design allowed the fuel and oxidizer to be very close together, which reduced the overall complexity of the rocket engine. *Mojave Aerospace Ventures LLC, photograph by David M. Moore*

The 22-inch- (56-centimeter-) diameter case of the CTN, where the combustion occurs, is essentially a hollow tube lined with the rubber fuel. The throat forced the hot exhaust gases, passing from the case to the nozzles, to converge. The nozzle accelerated the hot exhaust gases to provide even greater thrust. The silica/phenolic high-temperature insulating liner and mounting rings of the case were assembled together with the throat and nozzle using graphite/epoxy composite overwrap. The cantilevered mounting also provided the advantage of being able to vary the length and diameter of the CTN without design modifications.

A safety feature was built into the CTN to warn of an impending burn-through. If the fiber-optic cable sealed between the liner and the overwrap detected a breach in the liner, the flow of oxidizer would be cut off and the rocket engine shut down. As an additional precaution, wires wrapped around the outside of the CTN could also be used to detect a burn-through.

AAE Aerospace supplied an ablative nozzle with an expansion ratio of 25:1 for the CTN. The expansion ratio compares the size of the throat to the size of the nozzle's opening. The efficiency is altitude dependent, so this ratio gave *SpaceShipOne* the maximum thrust at launch altitude. The interior of the nozzle eroded away as the rocket engine burned, which helped to minimize the temperature of the nozzle itself.

The entire CTN had to be replaced after each spaceflight.

Scaled Composites looked to outside vendors to supply the remaining components for the rocket engine, which included the injector, the igniter, and the related controls and plumbing. Scaled Composites held a competition of its own to build these components.

Rocket-Engine Competition

Simplicity of design by reducing the number of parts is best illustrated by an experience Rutan had during his preliminary research into rocket-engine designs. Representatives of a rocket-engine vendor Rutan was visiting wanted to demonstrate how

Fig. 5.9. At the front of the oxidizer tank are the fill, vent, and dump valves. The vent valve protects from over-pressurization, and the dump valve allows the tank to be drained in an emergency situation. *Mojave Aerospace Ventures LLC, photograph by David M. Moore*

their valve worked after showing him schematics and describing their engine design. They went out to the back of the building, where the valve was chilled with liquid nitrogen to a cryogenic operating temperature.

"The valve stuck," Rutan said. "It didn't work. It failed right there while I was looking at it."

After they went back inside, he said to them, "Gentlemen, let me ask you, how many of those valves are in this motor?"

Twelve, they replied.

"Did you expect this to work today?" he asked.

There was no answer. Rutan knew enough about rocketry to know if there was such a thing as a simple one that operated at room temperature with one valve, that's the one he wanted.

Rutan had been looking at more than twenty rocket-engine vendors. With the help of Tim Pickens, a bit of a maverick rocket expert, Rutan sketched up the design and passed it on. "I had sent them a sanitized requirement that didn't divulge that we were doing manned flight. I said it was a reusable sounding rocket."

Now, it was necessary to build the remaining engine components. In front of the oxidizer tank was the fill, vent, and dump system as well as the forward tank bulkhead. This was how the liquid nitrous oxide was added to the tank, how pressure was relieved in case of over-pressurization, and how the nitrous oxide could be drained in an emergency situation. These components were kept away from the hot side of the oxidizer tank to improve safety. The locations of these components are shown in figure 5.9.

Arguably the most critical valve of the whole project was the main control valve that supplied the oxidizer to the CTN. Without the flow of nitrous oxide, there would be no rocket. So, this valve received special attention. Doug Shane cleverly named it the "oxy-no-more-on" valve. Connected to the oxy-no-more-on valve was the injector, which was essentially a showerhead that sprayed the liquid nitrous oxide into the CTN. Both the oxy-no-more-on valve and the injector were located inside the oxidizer tank. What this did was further reduce plumbing and had the added benefit of keeping these components at a constant temperature, which also simplified things. Figure 5.10 shows their positions from a rear view of the oxidizer tank.

The rear tank bulkhead, motor controller, head insulation, and igniter still needed to be built as well. The igniter was the heat source that initiated the reaction between the oxidizer and rubber fuel. The last part that remained was the actual rubber fuel itself, which had to be packed into the CTN.

From the list of more than twenty potential rocket-engine vendors, it came down to two: Environmental Aeroscience Corporation (eAc), of Miami, Florida, and SpaceDev, of Poway, California. This wasn't a low-bid process. It was performance-based, and the competition would go back and forth. Rutan thought that having just one vendor try to make the last of the components was too much of a risk. "When you sole source, if they stumble, you have to pay them to pick up," Rutan said. "If you've got two of them competing and they stumble, they spend their own money to catch up. That's what the big 'C'-word is—competition. And that happened. They both wanted to fly."

In the middle of 2001, both vendors began to participate in a ground-test program that included cold flow tests, where the oxidizer and fuel weren't ignited, as well as partial- and full-duration

Fig. 5.10. The CTN mounted on a titanium flange at the rear of the oxidizer tank. Located inside the tank was the main control valve, also known as the "oxy-no-more-on" valve, that released the oxidizer into the CTN. *Mojave Aerospace Ventures LLC, photograph by David M. Moore*

Fig. 5.11. The test stand trailer (TST) was used during the rocket-engine test firings. It contained all of the same components and systems to be used in *SpaceShipOne*, including an oxidizer tank that the CTNs were mounted onto. *Mojave Aerospace Ventures LLC, photograph by David M. Moore*

Fig. 5.12. To fill *SpaceShipOne* or the TST with the oxidizer, nitrous oxide (N_2O), Scaled Composites used a special tanker truck called the mobile nitrous oxide delivery system (MONODS). *Mojave Aerospace Ventures LLC, photograph by David M. Moore*

hot-firings. All rocket engine testing employed the same type of components used in *SpaceShipOne*. Shaking out the full-scale designs on the ground at a Scaled Composites test site rather than in flight was a much more preferable way to address the "ups"—break up, burn up, and blow up.

In terms of experience, eAc had built a lot of hybrid rocket engines and had run some with diameters of up to 16 inches (41 centimeters). SpaceDev, on the other hand, had only run little hybrids with 15 pounds-force (67 newtons) of thrust.

This meant eAc was off to a quick lead. By June of 2002, the first set of results came in. Scaled Composites selected eAc as the supplier of the front-end propulsion components. However, both vendors continued to develop the remaining rocket-engine components.

To run all the rocket-engine tests, the test stand trailer (TST), a partial, operational mockup of *SpaceShipOne* shown in figure 5.11, acted as a portable test bed. It comprised the complete rocket engine, including the CTN and oxidizer tank, and the forward section of the fuselage. The TST provided instrumentation to analyze vibration, temperature, and stress conditions that the flight components would face during spaceflight. It also had load cells, which were sensors mounted on the trailer, that allowed engineers to evaluate the rocket-engine performance.

Setup for rocket-engine tests typically took several days. Since the TST was mobile, preparations began inside a hangar at Scaled Composites. Once the CTN was in place and all the other components were hooked up, the TST was towed out to the firing range. Figure 5.12 shows the specialized tanker truck called the mobile nitrous oxide delivery system (MONODS) that then fills the oxidizer tank.

Mission Control ran the rocket-engine tests remotely, and an underground computer near the TST collected and transmitted live data. The countdown procedures for the ground testing are listed in table 5.1.

Since testing occurred on the ground, a nozzle with an expansion ratio of 10:1 was used, as opposed to 25:1 used for actual flights.

SpaceDev had fired first, a 15-second burn on November 21, 2002. The rocket roar was deafening for each hot-fire of the rocket engine. The ground shook. Figure 5.13 shows the tip of a fiery plume stretched out more than thirty feet, scorching the ground and sending smoke billowing across the desert.

"The SpaceDev motor was the only one that had a critical safety issue. In the first round, a valve wouldn't close and it keep burning and burning," Rutan said. Although Scaled Composites was slow to extinguish the fire, no significant damage occurred.

After more than nine months of firing rocket engines, the gap between eAc and SpaceDev had disappeared. Each had little differences. For example, SpaceDev used four ports to flow the oxidizer into the CTN, whereas eAc used just one, and they each had a proprietary rubber fuel and system for refueling the CTN. But both had exceeded performance expectations.

With the exception of the fire caused by a stuck valve at the very onset of test firing, the hybrid rocket engine demonstrated safety and reliability. The cantilever mounting concept proved structurally sound, the oxidizer and CTN performed as designed, and the "ups" were avoided altogether.

Scaled Composites had both companies commit prices for components, not only for the upcoming flight test and Ansari X Prize rocket engines, but for twenty-two other rocket engines slated for what was called Task 21, the spaceflights after the Ansari X Prize attempts.

Table 5.1 Rocket Engine Test Countdown
(source: Mojave Aerospace Ventures LLC, provided courtesy of Scaled Composites)

T-48 hours   Mounting the CTN
The fully fueled CTN (engine) and valve are mounted to the oxidizer tank using the mounting trolley. Alignment is checked and bolts are tightened to the specified tolerances.

T-36 hours   Instrumentation and Integration
The motor controller, pressure bottles, actuation valves, and all the sensors are installed and tested on the test bed. New CTNs require new temperature and strain sensors, while CTNs being fired for the second or third time are "plug and play."

T-24 hours   Moving Time
The test trailer with the fully fueled CTN (engine) is moved from Scaled's hangar to the test site. There it is bolted down, final instrumentation is completed, and the system is connected to Mission Control.

T-4 hours   Filling the Tank
Nitrous oxide is transferred from the MONODS to the oxidizer tank on the test trailer after it has been brought up to temperature and pressure. Temperatures and pressures must be carefully controlled to ensure a safe transfer.

T-0 hours   The Final Countdown
After final system checks have been made, cameras have been started, and range safety has been checked, a quick countdown is called and the switch is thrown. The motor controller automatically ignites and fires the engine for a preprogrammed period of time.

"SpaceDev's components were lighter, which helped our perform-ance," Rutan said. "Keep in mind, we knew that we had to have really the best performance in order to win the X Prize because we had a weight growth. So, we needed all the performance we could get. SpaceDev's components had a little more efficiency, on the order of three percent for the same weight of propellant."

SpaceDev actually turned out to be cheaper, too—all characteristics Rutan was very fond of. In September of 2003, the development phase of *SpaceShipOne*'s rocket engine was completed. SpaceDev got the nod.

Figure 5.14 shows a rocket engine installed in *SpaceShipOne* without the fairing in place, so the unsupported ended of the CTN is apparent.

Rocket-Engine Operation

On November 18, 2003, Scaled Composites qualified a rocket engine in preparation for the first powered flight of *SpaceShipOne*.

"The nice thing about this particular type of hybrid motor is the nitrous oxide is room temperature in the oxidizer tank in front. It self-pressurizes to about 750 pounds per square inch," Doug Shane said. This eliminated the need for complicated and expensive pumps to flow the oxidizer into the CTN. In figure 5.15 a panel on the belly of *SpaceShipOne* is removed, exposing the engine bay, so the CTN could be easily mounted.

The oxidizer tank was filled with liquid nitrous oxide through its forward bulkhead. The MONODS pumped the nitrous oxide at 300 pounds per square inch and 0 degrees Fahrenheit. To bring the temperature up to room temperature and maintain it there, engine bleed air from *White Knight* flowed into *SpaceShipOne* and was directed at both ends of the tank during captive carry.

The rocket engine of *SpaceShipOne* could only be fired once. After the rocket engine was armed and the fire switch thrown, the igniter produced a significant amount of heat, causing combustion between the oxidizer and fuel.

"*SpaceShipOne* does not have a throttle. It is on or off," Rutan said. "There was no need for a throttle."

For the spaceflights, the force of thrust was 16,800 pounds-force (74,730 newtons). However, because of performance concerns and

Fig. 5.13. Scaled Composites was able to monitor the performance of each rocket engine tested during the development phase. Data collected by sensors on the TST was relayed to engineers real-time for analysis. *Mojave Aerospace Ventures LLC, photograph by David M. Moore*

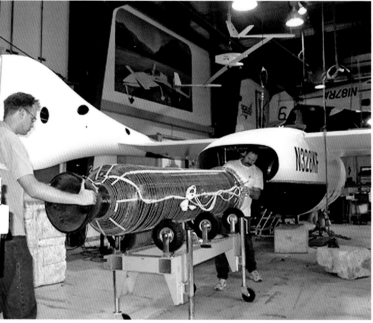

the additional payload required by the Ansari X Prize competition, additional nitrous oxide oxidizer and rubber fuel were added, as well as an increase in burn duration. The thrust and the specific impulse (I_{SP}), a measure of efficiency of the rocket engine, were not officially released by Scaled Composites.

Peak performance occurred once the rocket engine ignited. As the rubber burned away and the oxidizer tank emptied, the amount of thrust gradually reduced.

About sixty seconds into the burn, the nitrous oxide began a transition from liquid to gas because there was more room in the oxidizer tank for it to expand, which caused a significant decrease in thrust. At this point, the combustion was uneven throughout the CTN and caused *SpaceShipOne* to chug along for about five seconds. The ride then smoothed out for the remainder of the burn now that the nitrous oxide had transformed to gas.

During the flight, the engine performance and parameters could be monitored in the cockpit by the pilot and on the ground in Mission Control.

Brian Binnie stands in front of *SpaceShipOne* with its feather extended. In addition to the feather, *SpaceShipOne* required three other systems for control during a spaceflight: the mechanical flight control system for subsonic speeds, the electric flight control system for supersonic speeds, and the reaction control system (RCS) for space. *Mojave Aerospace Ventures LLC, photograph by Scaled Composites*

Chapter 6

SpaceShipOne
Instruments and Controls

*S*paceShipOne is part airplane, missile, spacecraft, and glider. Like the North American X-15, it was flown by hand. Figure 6.1 shows *SpaceShipOne* during a glide test flight with its control surfaces clearly visible on the tail booms.

To complete each spaceflight mission, it had to transition between a variety of flight phases, including takeoff, captive carry, boost, coast, feather, glide, approach, and landing. Flying these phases, *SpaceShipOne* crossed between three distinct flight regimes: subsonic, supersonic, and zero-g. Different flight control systems were required to operate in each regime. A basic mechanical control system was used in subsonic flight, while an electric-powered control system and a reaction control system (RCS) were required in supersonic and zero-g flight, respectively.

SpaceShipOne had a set of instruments and controls for each flight control system that the pilot had to switch between. Figure 6.2 shows Mike Melvill in the cockpit holding the control stick of *SpaceShipOne* with the nose cone detached.

Subsonic Flight Control

SpaceShipOne essentially flew like a light airplane—some of the time anyway. It had a mechanical flight control system that operated manually, similar to that of a Cessna 172 or Piper Cub. A simple cable-and-rod linkage tied the stick and rudder pedals

Fig. 6.1. *SpaceShipOne* was the first manned aircraft to use tail booms attached to the wingtips. Unlike almost every other aircraft, the wings had no control surfaces. The pilot maneuvered *SpaceShipOne* with control surfaces that were all located on the tail booms, an upper and lower rudder at the back of each tail boom, flying or movable horizontal stabilizers, and elevons at the trailing edge of each horizontal stabilizer. *Mojave Aerospace Ventures LLC, photograph by Scaled Composites*

Fig. 6.2. With the nose cone detached, Mike Melvill gets some practice at the controls of *SpaceShipOne* as Steve Losey, the crew chief, looks on. *SpaceShipOne* had room for a pilot up front and two passengers in the back row. *Mojave Aerospace Ventures LLC, photograph by David M. Moore*

to the control surfaces. The connections between the pilot's controls in the cockpit and the control surfaces are shown in figure 6.3. No hydraulics were used for flight control.

All of the control surfaces were located on the tail booms. An upper and lower rudder were mounted at the end of each tail boom. The rudder pedals in the cockpit moved the upper rudders for control of yaw. Refer to figure 6.4 for *SpaceShipOne*'s flight axes and rotations. Each rudder pedal worked independently and deflected the corresponding upper rudder outward only. By depressing both, the upper rudders acted a little bit like a speed brake.

On the horizontal stabilizer, elevons combined the functions of conventional ailerons on the wings and elevators on the tail. The control

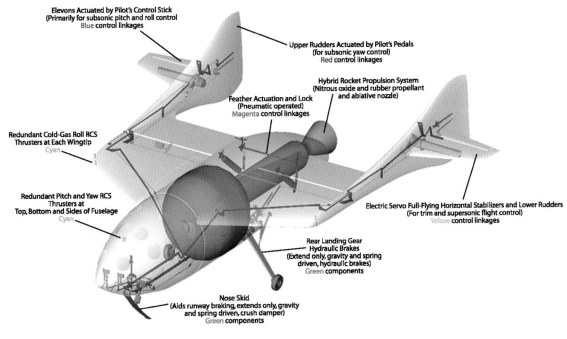

Elevons Actuated by Pilot's Control Stick
(Primarily for subsonic pitch and roll control
Blue control linkages

Upper Rudders Actuated by Pilot's Pedals
(for subsonic yaw control)
Red control linkages

Hybrid Rocket Propulsion System
(Nitrous oxide and rubber propellant
and abiative nozzle)

Feather Actuation and Lock
(Pneumatic operated)
Magenta control linkages

Redundant Cold-Gas Roll RCS
Thrusters at Each Wingtip
Cyan

Redundant Pitch and Yaw RCS
Thrusters at
Top, Bottom and Sides of Fuselage
Cyan

Electric Servo Full-Flying Horizontal Stabilizers and Lower Rudders
(For trim and supersonic flight control)
Yellow control linkages

Rear Landing Gear
Hydraulic Brakes
(Extend only, gravity and spring
driven, hydraulic brakes)
Green components

Nose Skid
(Aids runway braking, extends only, gravity
and spring driven, crush damper)
Green components

Fig. 6.3. This diagram shows the different control systems and the linkages from the controls in the cockpit to the control surfaces. The reaction control system (RCS) and the landing gear are also shown in this cutaway. *Mojave Aerospace Ventures LLC, provided courtesy of Scaled Composites*

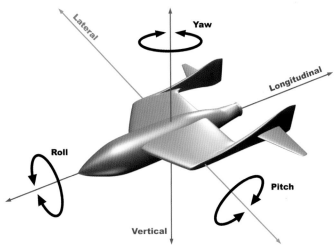

Lateral

Yaw

Longitudinal

Roll

Pitch

Vertical

Fig. 6.4. Each of the three flight control systems allowed *SpaceShipOne* to maneuver along the flight axes. Pitch and roll were controlled by the elevons at subsonic speeds and by the movable horizontal stabilizers at supersonic speeds. The upper rudders and lower rudders controlled yaw at subsonic and supersonic speeds, respectively. RCS thrusters mounted on the wings and fuselage controlled pitch, roll, and yaw in space. *James Linehan*

Fig. 6.5. At the trailing edge of each horizontal stabilizer was an elevon. When both elevons moved in the same direction, the pitch of *SpaceShipOne* changed. When moved differentially, roll changed. The photograph shows the poor air flow over a horizontal stabilizer during a stall test, as indicated by the small strings attached to the surface. *Mojave Aerospace Ventures LLC, video capture provided courtesy of Discovery Channel and Vulcan Productions, Inc.*

stick moved the elevons, which allowed the pilot to control pitch and roll of *SpaceShipOne*. Figure 6.5 shows a horizontal stabilizer during testing after modification.

Supersonic Flight Control
Similar to the Bell X-1, as shown in figure 6.6, which in 1947 was the first aircraft to break the sound barrier, *SpaceShipOne* used electric, motor-driven control surfaces to maneuver. After *SpaceShipOne*

broke the speed of sound during test flights, the subsonic flight controls no longer functioned efficiently.

"Once you are supersonic, the control system that you normally use to fly the airplane doesn't work anymore," Melvill said. "Because it is a mechanically controlled airplane, there is no hydraulic system like there is on an F-16 or F-18. It is just cables and pushrods. You are just not strong enough to move the controls at that point. So you just revert to using the trim switches."

Fig. 6.6. Because the forces pushing against the control surfaces of the X-1 were so strong while flying at supersonic speeds, the pilot could not use conventional mechanically linked flight controls. The control surfaces had to be electrically controlled and moved using electric motors. *SpaceShipOne* used a similar system for flying above Mach 1. *NASA–Dryden Flight Research Center*

Many aircraft have trim controls that help the pilots maintain course by slightly altering control surfaces. By changing a trim setting, a pilot can fine-tune the control surfaces, so less force has to be applied to the control stick and rudder pedals in order to stay on course. *SpaceShipOne* uses trim controls in a somewhat similar way during supersonic flight.

A switch on the top of the control stick activated electric servos that pivoted the entire horizontal stabilizer on each side of *SpaceShipOne* for pitch trim. In figure 6.7, a close-up of a tail boom shows the numeric scale that indicates the amount of deflection for the horizontal stabilizer as set by the pilot.

Fig. 6.7. Triangular strakes were added in front of the horizontal stabilizers, which were also enlarged, to improve the aerodynamics of *SpaceShipOne*. The numbers to the right of the strake show the amount of trim or deflection of moveable horizontal stabilizers. *Mojave Aerospace Ventures LLC, photograph by David M. Moore*

A knob at the pilot's left side called the "turtle" activated the lower half of each rudder, giving yaw trim. "Yawing the airplane caused it to roll because of the high wing and the swept leading edges," Melvill said. "We actually didn't use the roll trim. It was too powerful. We used yaw trim to roll the plane. If it was rolling off to the left, you would yaw it to the right."

The lower rudders were synced and moved in the same direction, unlike the upper rudders used in subsonic flight. "The geometry is such that they go out a lot and in just a little bit," Rutan added. "But they never work opposite."

Trimming was also used to restrict the movement of *SpaceShipOne* during test flights. For example, when Mission Control monitored the trajectory of *SpaceShipOne*, they would instruct the pilot to make trim changes in order to help him stay on course.

More than midway through the burn, the atmosphere became so thin that the supersonic flight controls were no longer needed. The pilot was able to control *SpaceShipOne* with subsonic flight controls, even though it was flying much faster than the speed of sound, because the rarified air produced little opposing force. When *SpaceShipOne* fully left the atmosphere, the pilot then switched over to the RCS.

Reaction Control System

Because there is no atmosphere in space, the flight control systems that ordinarily allow an aircraft to move through the air do not work for spacecraft moving through space. Rudders, elevators, and ailerons only work because air moves over them. With no air, they are useless.

In order to maneuver in space, spacecraft take advantage of a simple physics law discovered by Sir Isaac Newton: *for every action, there is an equal and opposite reaction*.

Without considering a spacesuit, for example, if a person was in space and blew through a straw, the air would move out the straw in one direction and the person would move in the opposite direction.

Figure 6.8 shows an astronaut with a hand-held reaction control system (RCS). To move, he just points the opposite direction, releases a puff of gas, and off he goes in the direction he wants. This, by the way, is the same principle by which a rocket engine works. The RCS thrusters are just miniature rocket engines.

Fig. 6.8. In Gemini 4, astronaut Ed White made the first U.S. spacewalk. To maneuver during his 23-minute extravehicular activity (EVA), he used a hand-held self maneuvering unit (HHSMU) that shot little bursts of gas, which allowed him to move around. This device worked similar to the way *SpaceShipOne*'s reaction control system works. *NASA–Johnson Space Center*

The Space Shuttle uses a fuel of monomethylhydrazine (MMH) and an oxidizer of nitrogen tetroxide (N_2O_4) for its RCS. These propellants react together spontaneously once in contact. As long as each chemical is stored safely separate, they provide the orbiter a simple, reliable, precise, and powerful RCS.

SpaceShipOne had limited time in space and was much less massive than the Space Shuttle. The force generated by these expensive and toxic chemicals was not required. So, puffs of air were sufficient to maneuver *SpaceShipOne* in space.

After the aerodynamic control authority was gone, the pilot used the RCS to help slow down or null any rotation that had developed while exiting the atmosphere. Each wingtip had roll thrusters, and along the top, bottom, and sides of the fuselage were pitch and yaw thrusters. Each of these thrusters was essentially a port from which high-pressure air could be expelled, and each thruster had a backup. Redundant 6,000-psi bottles of air powered the RCS. By fully extending the rudder petals and the control stick, the pilot maneuvered *SpaceShipOne* by triggering microswitches that turned the appropriate thrusters either full-on or full-off.

The RCS was also used to get into position for reentry. Scaled Composites had confidence that the feather would self-right *SpaceShipOne*. However, they did not want to start off upside-down if they didn't have to.

Tier One Navigation Unit

The pilot had to fly a specific trajectory carefully during a mission. If he deviated, he risked not only failing to reach the target altitude but also missing the prescribed reentry area or, in the extreme case, being too far away from the landing site.

"The aircraft itself was completely manually controlled," Pete Siebold said. "So, the only feedback the pilot had to how the airplane was flying was through the avionics system."

It was necessary to develop an avionics system, called the Tier One navigation unit (TONU), for *SpaceShipOne*. "There really was nothing available within our budget and nothing available off the shelf that suited our needs. So, we had to go develop it ourselves," Siebold said. The system navigation unit (SNU) and the flight director display (FDD) were the two primary components that made up the TONU.

"We had contracted a company to basically develop the hardware portion of the nav system," Siebold said. "They built the boxes and put the computers in. They were initially responsible for developing the software of the navigation system as well. However, we ended up making major modifications to that software at the end of the program to make it perform the way we needed it to perform. On the display side, we wrote all the software for the entire program from the beginning." Aside from being a test pilot, Siebold was the engineer behind most of the software design. Fundamental Technology Systems (FTS), also an Ansari X Prize competitor, provided the hardware and initial software to Scaled Composites.

Acting as the brain of the TONU, the SNU incorporated both a global positioning system (GPS) and an inertial navigation system (INS). It sent guidance and navigational information to the pilot, who saw it on the liquid crystal display (LCD) screen of the FDD in glass-cockpit-type fashion. The SNU navigates along the primary flight axes in six degrees of freedom: the translations of left/right, forward/back, and up/down and the rotations of yaw, roll, and pitch.

Close-ups of the FDD are shown in figure 6.9 and figure 6.10, which also show the similarity between the cockpits of *SpaceShipOne* and *White Knight*.

Fly-by-wire was not an option. Siebold said, "It wasn't warranted for the complexity of this program. Fly-by-wire adds a whole order of magnitude to the whole vehicle development costs. And we really wanted to keep this as simple as possible in order to make this affordable for everybody. That is really the backbone of this program. If you can make it as simple as a Volkswagen, then everybody can afford it. If it needs to be as complex as the Space Shuttle, then nobody can afford it. We really had to push really hard toward making it affordable from the onset."

The data available to the pilot is based on several modes that correspond to the different phases of flight for *SpaceShipOne*. Figures 6.11 to 6.14 show various FDD modes, including a boost, a reentry, and a glide. In these modes, the pilot is given trajectory guidance with respect to a detailed map that tracks the position of *SpaceShipOne*. The FDD automatically stepped through the different modes while flying the mission, but a control allowed the pilot to manually move through the modes in the unlikely event he needed to do so.

Siebold said, "We had the initial boost portion. So, that was the pull up. Then it transitioned to a pseudo-boost mode were everything zoomed in and allowed you to track your final target, fly that

Fig. 6.9. A close-up of the flight director display (FDD) for *SpaceShipOne* is shown in this photograph. The FDD was part of the Tier One navigation unit (TONU) and provided the test pilot with instruments similar to the way a glass cockpit does for an airliner. *Mojave Aerospace Ventures LLC, photograph by David M. Moore*

Fig. 6.10. The inside of *White Knight*'s cockpit, shown here, is remarkably similar to the cockpit of *SpaceShipOne*. Even the instrumentation and controls are nearly identical, with the obvious exception that *SpaceShipOne* has rocket-engine controls and *White Knight* has jet-engine controls. Since *White Knight* started flying about a year before *SpaceShipOne*, this allowed Scaled Composites to build up confidence in the instruments prior to flying *SpaceShipOne*. Also, *White Knight* could be used as a trainer for *SpaceShipOne*. *Mojave Aerospace Ventures LLC, photograph by Scaled Composites*

apple onto your target. Once the motor shut down, it transitioned to a coast phase. Once you left the atmosphere, it transitioned into a reentry phase. Once you reentered, it transitioned into three different glide phases. We called them high key, final, and landing phase. And those three phases helped you to find your way back to the airport, and manage your energy so that you'd end up touching down at the place you wanted."

Sometimes *SpaceShipOne* nearly pointed straight up, and sometimes it was upside-down. The attitude, or orientation, of *SpaceShipOne* in flight was key flight information provided on the FDD. "It showed you whether or not you are at wings level," Siebold explained. "One unique aspect of the display was that as you pitched the nose up, when the horizon on the display disappeared, it still gave you situational feedback to tell you what attitude the aircraft was in."

A second key piece of flight information was the velocity vector. "What that tells you," Siebold said, "is the direction in which you are currently flying—the direction in which your velocity is currently heading. That was depicted on the screen with what we called the green apple. It was a green circle with a tail and two wings pointing out of it."

So, with the pilot knowing how *SpaceShipOne* was oriented and how it was moving in flight, the FDD offered two other bits of crucial flight information. These were the location of the optimum trajectory, represented by the "red donut," and where *SpaceShipOne* was with respect to it, which was the "green apple."

In a presentation at NASA Ames, Doug Shane had given the following succinct description: "The goal is to take that green velocity vector and put it right over that red donut, because that is the flight-director cue. And that gets you to the reentry point that you want. Very simplemindedly, your task is only to get those two circles closed up as quickly as you can. And that establishes essentially a vertical trajectory and gives you the best performance that you can get."

In addition, the SNU monitored and recorded how the systems of *SpaceShipOne* were performing and fed this information to the FDD, where it was displayed. "It acted as a caution/warning/advisory system," Siebold said. "It told you if there is any parameter out of limits, or if

there is anything out of limits that would cause you to abort the flight. We had a small list of parameters that if they ever exceeded some allowable range, they'd flash a big red sign that said 'abort.'"

During the rocket engine burn, things happened fast. There was not a lot of time to make decisions. The TONU did not automatically control *SpaceShipOne*. "The pilot still had to look at the information, digest it, and make the appropriate decision with that information," Siebold said.

Data that the SNU collected then displayed to the pilot on the FDD was also transmitted real-time to Mission Control on the ground by a radio frequency (RF) telemetry downlink. In Mission Control, the data reduction system (DRS) collected, processed, and stored all the transmitted data and made it accessible to everyone in Mission Control.

Energy Altitude Predictor

During the test flight, the TONU displayed a readout from the energy altitude predictor during the boost phase. Developed by aerodynamicist Jim Tighe, it worked by making calculations based on factors like *SpaceShipOne*'s speed and thrust. The pilot used this to decide when to turn off the rocket engine because *SpaceShipOne* was roughly half the distance to apogee after rocket-engine shutdown. For the second half, it coasted the rest of the way up.

The pilot needed a way to ensure he didn't run the rocket engine too long or too short. The initial powered flights relied on a timer, but using the energy altitude predictor yielded much better results. By looking at the readout of the energy altitude predictor, the pilot had a very good idea of the altitude *SpaceShipOne* would reach if the rocket engine shut down at that exact moment.

For example, the energy altitude predictor may have read 200,000 feet (60,960 meters), but in actuality, *SpaceShipOne* may have only been at an altitude of 80,000 feet (24,380 meters). So, if

SpaceShipOne shut down the rocket engine at that precise moment, it would coast to an apogee of about 200,000 feet (60,960 meters). This would be 128,000 feet (39,010 meters) short of the Ansari X Prize requirement. Therefore, the pilot wouldn't have shut down the rocket engine at this point, but he would have waited until the energy altitude predictor read at least 328,000 feet (100,000 meters).

Cockpit Instrumentation

Inside the cockpit, mixed between the circular windows, ports, door, and escape hatch, the pilot had all the instruments and controls he needed for all the various phases of flight. Figures 6.15 to 6.17 show views of the instruments and controls at the front of the cockpit, to the left side of the pilot, and to the right side of the pilot, respectively. The instruments and controls are identified by their numbered callouts given from these figures.

The pilot used the control stick (33) and the rudder pedals (7 and 12) to fly subsonically and to maneuver using the RCS (15, 38, and 41).

But for supersonic flight, trims (31, 35, and 36) and backups (55) were used. The controls for the rocket engine included switches and a timer (34, 42, and 49–52). The feather was operated using valves and levers (1, 3, 39, 40, 43, and 44).

Cameras on the tail, on the fuselage, and in the cockpit (56) provided video that was also an important source of data during flight testing. Mission Control used one of the cameras to monitor the feather and rocket engine in real-time. Two of the more unexpected things found in the cockpit were the ping-pong ball (8), which was used to provide a good visual during weightlessness, and the "Q-tip" (63), which the pilot used to wipe down excess moisture from the windows.

The controls for the ECS (13, 16, 17, and 64–69), battery (18–22), landing gear (45 and 53), radios (30), and other systems were also close at hand. However, important instruments like the airspeed indicator, Machmeter, altimeter, and energy altitude predictor were all displayed on the FDD of the TONU (11), and airspeed and altitude were also backed up on the Dynon (10).

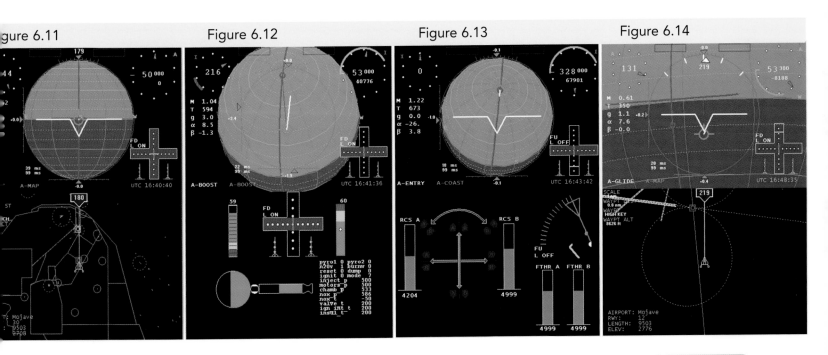

Figure 6.11   Figure 6.12   Figure 6.13   Figure 6.14

Fig. 6.11. One of the components of the Tier One navigation unit (TONU) was the flight director display (FDD). Like the glass cockpit of an airliner, the FDD showed many of the important instruments and readouts used by the pilot to fly *SpaceShipOne*. An initialize mode of the FDD is shown with *SpaceShipOne* lined up on Runway 30 of Mojave Airport. *Mojave Aerospace Ventures LLC, provided courtesy of Scaled Composites*

Fig. 6.12. As *SpaceShipOne* rockets to space, a boost mode is shown on the FDD. By closing together the red circle and green circle, the pilot achieved optimum trajectory. The pilot could also view the status of the rocket engine and oxidizer tank. *Mojave Aerospace Ventures LLC, provided courtesy of Scaled Composites*

Fig. 6.13. The FDD shows a reentry mode before *SpaceShipOne* returns to Earth's atmosphere. The position of the feather, the operation of the reaction control system (RCS), and the condition of their pressurization sources are displayed. *Mojave Aerospace Ventures LLC, provided courtesy of Scaled Composites*

Fig. 6.14. After reentry and the feather is retracted, *SpaceShipOne* glides back to Mojave Airport. The test pilot used a glide mode on the FDD to help ensure *SpaceShipOne* reached the runway at correct position and speed. *Mojave Aerospace Ventures LLC, provided courtesy of Scaled Composites*

Figure 6.15

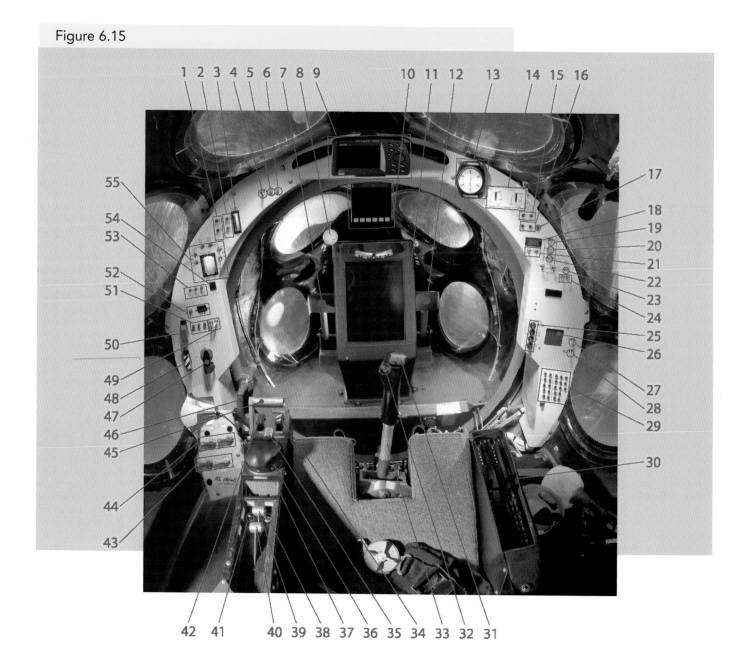

1. Feather bottle low: A and B
2. Wing against stops and wing TE locked down indicators
3. Feather position
4. Launch separation controller
5. Spaceship "Armed" indicator
6. Mothership "Armed" indicator
7. Left rudder pedal
8. Ping-pong ball
9. Backup GPS navigation display
10. Dynon backup altitude indicator
11. FDD (flight director display) of the TONU (Tier One navigation unit)
12. Right rudder pedal
13. Cabin altitude gauge

14. Landing pattern attitudes: normal and emergency (gear down)
15. RCS bottle pressure warning lights: A and B
16. ECS bottle pressure warning lights: A and B
17. Cabin pressure low warning light
18. Battery voltage
19. Bus tie
20. A battery
21. Selector switches
22. B battery
23. Video transmit power
24. TONU power
25. Trim circuit breakers: left stabilizer, right stabilizer, yaw, and backup trim
26. Backup rate display

27. Stabilizer boost
28. Damper heat
29. Circuit breaker panel indicators
30. Communication/navigation panel: two radios, transponder, and intercom selector panel
31. Pitch trim
32. Red button not used
33. Pilot roll/pitch control stick
34. Rocket motor fire
35. Roll trim
36. Yaw trim
37. FDD page control switches

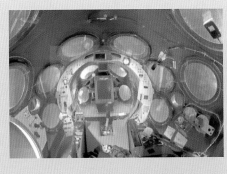

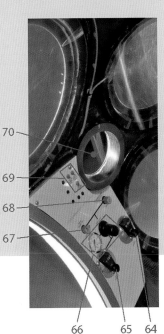

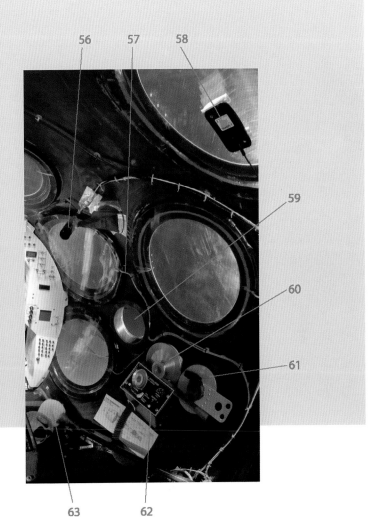

38. RCS A enable switches
39. Feather actuator
40. Feather unlock
41. RCS B enable switches
42. Rocket motor arm
43. Feather lock pressure valves: A and B (the yellow feather lock valve also doubles as gear down emergency assist)
44. Feather actuator pressure valves: A and B
45. Landing gear handle
46. Nose cone release handle
47. Nitrous oxide dump valve
48. Backup dump (through main valve)
49. Rocket motor controller power: A and B buses
50. Rocket motor controller reset
51. Motor armed indicator, main oxidizer valve commanded open indicator, and nitrogen pressure low indicator
52. Rocket motor burn time controller
53. Landing gear down indicators: left, nose, and right
54. Lamp test
55. Backup trim (stabilizer) panel

56. Lipstick camera, forward cabin (focused on pilot)
57. Dry air feed line (vents between window panes to prevent fogging)
58. GPS antenna (attached to window)
59. 4-inch opening for fine cabin pressure relief valve
60. Fine cabin pressure relief valve (this is the storage location)
61. Emergency cabin pressure dump port (this is the storage location)
62. Oxygen control panel
63. "Q-tip"

64. Secondary cabin pressure bottle valve
65. Primary cabin pressure bottle valve
66. Pressure regulator and gauge
67. Defog control valve (for between window panes)
68. Cabin make-up air
69. Dehumidifier fans and CO_2 scrubber fan switches
70. 4-inch opening for emergency cabin pressure dump port

photo by Eric Long and Mark Avion, National Air and Space Museum, Smithsonian Institution

After construction of *SpaceShipOne* and flight testing of *White Knight* were completed, it was time to begin flight testing *SpaceShipOne*. Flight tests consisted of captive carries, unpowered glides, rocket-powered flights, and spaceflights. *Mojave Aerospace Ventures LLC, photograph by Scaled Composites*

Chapter 7

Test Flights Begin

As soon as *SpaceShipOne* was able to fly, flight testing began. This didn't mean it was fully operational—not by a long shot. But the only way to learn more about the spacecraft was to get it up in the air and see how things worked.

The purpose of flight testing was to understand how the parts of *SpaceShipOne* operated individually and collectively while in flight. Only so much can be evaluated by ground tests and computer models. *SpaceShipOne* was, after all, designed to fly, not to take up idle space inside a hangar.

SpaceShipOne's flight test program proceeded incrementally, as with most flight test programs. After each step, Scaled Composites knew a little bit more how *SpaceShipOne* flew. Resolving problems, expected and unexpected, by making modifications or procedural changes is a normal part of any flight test program. And this one was no different.

SpaceShipOne completed fifteen test flights before attempting to capture the Ansari X Prize: three captive carries, eight glides, and four powered flights. These test flights, beginning on May 20, 2003, enabled the team to qualify the instruments, controls, systems, and test pilots as well as expand the flight envelope of *SpaceShipOne*.

Flight testing of *SpaceShipOne* actually began in *White Knight*. Since the cockpits were virtually identical, as were many of the controls and components, *SpaceShipOne* had a tremendous head start by the time it finally reached the sky.

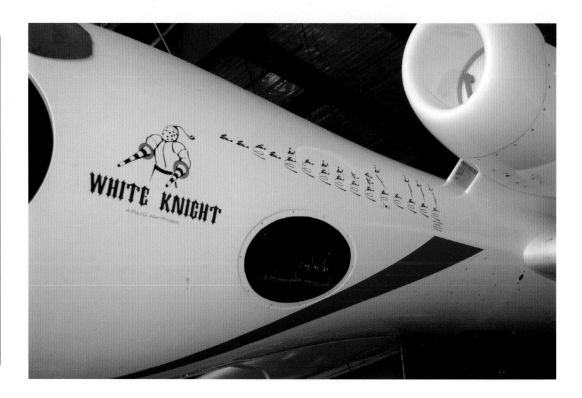

Fig. 7.1. *White Knight* flew more than four times the number of flights as did *SpaceShipOne* because it was also used to shake-out the components that were common to *SpaceShipOne*. Because they shared very similar cockpits and some similar flight characteristics, *White Knight* was also an effective flight trainer for *SpaceShipOne*. When *White Knight* and *SpaceShipOne* did fly together, a "cave painting" was added to the fuselage of *White Knight* to acknowledge each mission. *Mojave Aerospace Ventures LLC, photograph by Scaled Composites*

In this chapter and the following chapters, excerpts taken from the flight log of Scaled Composites are given for all the flights made by *SpaceShipOne* during the flight test program, including the Ansari X Prize attempts. The key information given for each includes the date, the flight numbers for both *SpaceShipOne* and *White Knight*, the crew members, and the test flight objective. From the photographs, the sequence of modifications to the tail booms, engine fairing, thermal protection system, and paint scheme becomes apparent.

For both *SpaceShipOne* and *White Knight*, the flight numbers were sequential, representing each time the vehicle flew. *White Knight* flew many more times because it was flown first and also used as a trainer for *SpaceShipOne*. For captive-carry test flights, a *C* was added to the end. For a glide or powered test flight, a *G* or *P* was added, respectively, while *L* denoted a launch for *White Knight*. Appendix A lists important data about each of the flights.

Piloting rotated between Pete Siebold, Mike Melvill, and Brian Binnie. But while one of these test pilots flew *SpaceShipOne* during a test flight, one of the others always flew *White Knight*. The remaining of the three test pilots typically flew a chase plane to closely observe the test flight. Appendix B gives details about the chase planes for the flights.

To commemorate the completion of each *SpaceShipOne* flight, a "cave painting" was added to the side of *White Knight*, as shown in figure 7.1.

Being a test pilot is a risky occupation, as history has continually proven. Anytime a new aircraft is flown, there are many unknowns. The test pilot has to be prepared for the worst case, though. The nose cone of *SpaceShipOne* was specially designed to detach from the fuselage, so, in an emergency, the test pilot could make a quick escape and parachute to safety. The job of the test pilots was to fly *SpaceShipOne* and, with the help of the entire test flight team, turn those unknowns into knowns.

Flight Test Log Excerpt for 1C

Date: 20 May 2003

	Flight Number	Pilot/Flight Engineer
SpaceShipOne	1C	none
White Knight	24C	Pete Siebold/Brian Binnie

Objective: First captive-carry flight with mated *White Knight* and *SpaceShipOne*. Vibration and aerodynamic interface assessment. Mated handling qualities evaluation. Envelope expansion to 130 knots/Mach 0.5 above 45,000 feet [13,720 meters]. Stalls and 2/3-rudder sideslips. *SpaceShipOne* systems inactive, controls locked, and cabin unmanned. Launch system was qualified and functional for this flight.

(source: Mojave Aerospace Ventures LLC, provided courtesy of Scaled Composites)

First Flight Test (1C)

Thirty-two days after Scaled Composites revealed its space program to the world, *SpaceShipOne* was about to take to the air for the first time. "Here we are about to embark on a flight test program with a spaceship," said Doug Shane, the director of the flight test program. "And we started off with an unmanned captive-carry flight just to make sure the interactions between the two airplanes were fine."

The first test flights off the ground were captive carries, where *White Knight* and *SpaceShipOne* took off attached together in the mated configuration and did not separate during the flight. This was just like a giant wind tunnel, but instead of an enormous fan blowing air on *SpaceShipOne*, *White Knight* moved *SpaceShipOne* through the

Fig. 7.2. *SpaceShipOne* took to the air for the first time on May 20, 2003, for an unmanned captive carry. A primary goal of the flight was to ensure that the two vehicles could safely fly while mated. *Mojave Aerospace Ventures LLC, photograph by Scaled Composites*

air. Figure 7.2 shows *SpaceShipOne* and *White Knight* during the first captive carry.

With Pete Siebold at the controls and Brian Binnie in its backseat as flight engineer, *White Knight* carried the unmanned *SpaceShipOne* to an altitude of 48,000 feet (14,630 meters), which would eventually be the approximate launch altitude. They reached a speed of Mach 0.53 after the 700 feet per minute (210 meters per minute) climb to this altitude. Siebold and Binnie flew for 1.8 hours and found that *White Knight* had excellent handling qualities and could perform the captive carry without any stability, interference, or vibration problems. *SpaceShipOne* was now ready for a pilot.

SpaceShipOne Now Manned (2C)

Within specifically designated airspace, Mike Melvill rode inside *SpaceShipOne* during a 2.1-hour-long captive-carry test flight. Aside

Flight Test Log Excerpt for 2C

Date: 29 July 2003

	Flight Number	Pilot/Flight Engineer
SpaceShipOne	2C	Mike Melvill
White Knight	29C	Brian Binnie/Cory Bird

Objective: First manned captive-carry flight of *SpaceShipOne*. A man-in-loop launch rehearsal and inflight checkout of all ship systems, including flight controls and propulsion system plumbing.

(source: Mojave Aerospace Ventures LLC, provided courtesy of Scaled Composites)

Fig. 7.3. Two months after *SpaceShipOne*'s first test flight, Mike Melvill became the first test pilot to get behind the stick of *SpaceShipOne*. This mission was a captive carry, though. The Starship, designed by Burt Rutan, flew as a chase plane. *Mojave Aerospace Ventures LLC, photograph by Scaled Composites*

Flight Test Log Excerpt for 3G

Date: 7 August 2003

	Flight Number	Pilot/Flight Engineer
SpaceShipOne	3G	Mike Melvill
White Knight	30L	Brian Binnie/Cory Bird

Objective: First glide flight of *SpaceShipOne*.

(source: Mojave Aerospace Ventures LLC, provided courtesy of Scaled Composites)

from checking systems, Melvill preformed a full rehearsal for the first glide-test flight with Brian Binnie and Cory Bird, who crewed *White Knight*.

Figure 7.3 shows this test flight and the Starship chase plane trailing behind. Several chase planes flew alongside *SpaceShipOne* and *White Knight* at various stages during the flight test program. They monitored how *SpaceShipOne* performed from an external standpoint, and, should *SpaceShipOne* or *White Knight* run into difficulties, they provided a valuable extra set of eyes.

During this rehearsal, *SpaceShipOne* also practiced the communication that would take place, which included sending data and video down to Mission Control on the ground.

Although pilots trained to fly *SpaceShipOne* with the simulator and *White Knight*, this was the first time that a pilot could actually feel the forces on the controls during flight. After Melvill exercised all the different systems aboard *SpaceShipOne*, which included parts of the feather and propulsion systems, *SpaceShipOne* was ready for the big next step.

As *White Knight* came in for a landing, though, Melvill couldn't help trying to land, even though *SpaceShipOne* still remained fixed to *White Knight*. From the chase plane video, *SpaceShipOne*'s elevons moved as *White Knight* flared for landing. Afterward, Melvill joked with Binnie by congratulating him on such a fine landing.

Going it Alone (3G)

"What a great thing to be able to fly glide selections without worrying about the rocket propulsion system or any of the other elements. And by separating all these variables out, you can learn how to fly the airplane and make sure all the subsonic stuff is going to work," Doug Shane said. In theory and in practice, yes, but it was still unnerving from a test pilot's point of view when it came down to a vehicle that had never been flown before.

"The first glide flight was probably the one that was my least favorite because we didn't even know if it would fly," Mike Melvill said. "If you think about a normal airplane with an engine, we don't just go out and fly it. We go out and taxi it slowly. We figure out if the brakes work, whether the steering works, and then we go a little bit faster until we can finally lift it off a few inches and say, 'Yeah, looks like its going to fly.' And then we fly.

"For this one, we just hooked it on the bottom of *White Knight*, went to about 50,000 feet [15,240 meters], and dropped it off. So, we didn't have a clue how it would fly or whether it would be good, bad, or indifferent. It wasn't great. We had to modify it a little bit. But it was flyable, and I was able to bring it back. But that was the scariest flight, I think. We just didn't have any knowledge other than Burt's 'That looks about right.' It had never been in a wind tunnel. There was no formal wind-tunnel testing of the airplane at all. And so we tested it in the real wind tunnel."

White Knight flying at 105 knots, 12 miles (19 kilometers) east of Mojave, released *SpaceShipOne* at an altitude of 47,000 feet (14,330 meters). The spaceship and mothership separated cleanly. *SpaceShipOne* flew freely for the first time and was stable once disconnected.

Over the 19-minute flight time, Melvill evaluated the handling and performance. During that short time, the controls and avionics operated as expected while he began to expand the flight envelope. At Mojave Airport on Runway 30, *SpaceShipOne* came in nice and easy to make its first landing.

Flight Aborted (4GC)

The fourth flight of *SpaceShipOne* was scheduled to be a glide flight. Figure 7.4 shows preparations being made the day before the flight. Melvill had successfully opened more than 60 percent of the subsonic flight envelope on the previous flight. This included speeds from stall to 150 knots. Now it was planned to open up the envelope even further.

The flight envelope indicated all the ways *SpaceShipOne* could safely fly, which included variations and combinations of speed, altitude, attitude, and other factors. The crews flew the easiest stuff, as the pilots and engineers gained confidence and began to nail down the flying characteristics. Step by step, they pushed the boundaries as they flew the vehicle more aggressively to reveal its limitations.

However, twenty minutes prior to separation, the launch had to be aborted due to a GPS malfunction with the avionics system. Figure 7.5 shows the mated pair during the aborted test flight prior to landing. Initially designated as 4G and 31L for *SpaceShipOne* and *White Knight*, respectively, the flight numbers were modified to 4GC and 31LC as a result of the abort. But it was possible to complete some systems testing prior to landing.

Flight Test Log Excerpt for 4GC

Date: 27 August 2003

	Flight Number	Pilot/Flight Engineer
SpaceShipOne	4GC	Mike Melvill
White Knight	31LC	Brian Binnie/Cory Bird

Objective: Second glide flight of *SpaceShipOne*. Flying qualities and performance in the spaceship feather mode. Pilot workload and situational awareness while transitioning and handling qualities assessment when reconfigured. As a glider, deep stall investigation both at high and low altitude and envelope expansion out to 200 knots and 4 g's. Lateral directional characteristics including adverse yaw, roll rate effectiveness and control, including aileron roll and full rudder sideslips.

(source: Mojave Aerospace Ventures LLC, provided courtesy of Scaled Composites)

Fig. 7.4. *SpaceShipOne* and *White Knight* are poised before the fourth test flight. Planned as *SpaceShipOne*'s second glide flight, the flight had to be aborted midair. *Mojave Aerospace Ventures LLC, photograph by David M. Moore*

Fig. 7.5. It was necessary to abort the fourth test flight because of an avionics malfunction involving the global positioning system (GP[ ] t separation. However, before returning to Mojave, test pilot Mike Melvill completed some systems testing. *Mojave Aerospace Ventures LLC, photograph by David M. Moore*

Feather Up (5G)

After resolving the avionics malfunction that caused the aborted glide flight, *SpaceShipOne* and *White Knight* were back up flying again the very same day. Melvill was dropped at 48,200 feet (14,690 meters) from *White Knight* flying at a speed of 105 knots. For his first maneuver, he put *SpaceShipOne* into a full stall to investigate stall characteristics.

The second maneuver was one of the most critical firsts of the entire flight test program. Evaluation of the feather would begin on this flight. The purpose of the feather was to decelerate *SpaceShipOne* during reentry into the atmosphere.

"That's something you do in glide tests," Burt Rutan said. "You don't have to do that in spaceflight because once you decelerate from your spaceflight, you find yourself in a stable glide, which is identical to the way we flew the airplane on its first glide flight. So, we went out early in the program and put the feather up and put it down."

Rutan had planned to do a high-speed pull-up in a glide flight and put the feather up as it peaked to simulate zero-g during the beginning of the program. But this turned out unnecessary and would have used up too much altitude. "We started off at 43,000 feet [13,110 meters] and put the feather up to make sure it flew the way we wanted," Doug Shane said. "We ended up doing feather

Flight Test Log Excerpt for 5G

Date: 27 August 2003

	Flight Number	Pilot/Flight Engineer
SpaceShipOne	5G	Mike Melvill
White Knight	32L	Brian Binnie/Cory Bird

Objective: Same objectives as the aborted flight 31LC/4GC earlier today. Second glide flight of *SpaceShipOne*. Flyi[ ] qualities and performance in the spaceship reentry or "feather" mode. Pilot workload and situational awareness while transitioning and handling qualities assessment when reconfigured. As a glider, stall investigation both at high and low altitude and envelope expansion out to 200 knots and 4 g's. More aggressive, lateral directional characteristics including adverse yaw, roll rate effectiveness and control, including 360 degrees aileron roll, and full rudder side slips.

(source: Mojave Aerospace Ventures LLC, provided courtesy of Scaled Composites)

deployments from tail-slide entries, and it just worked great. Everything was as good as we could have possibly hoped for."

SpaceShipOne was gliding along at an airspeed of 90 knots when Melvill unlocked and activated the feather. As the tail booms began to elevate to their fully extended position of 65 degrees, the nose of *SpaceShipOne* pitched up but settled back to a near-level pitch. Melvill encountered a lot of buzzing and buffeting during the 70-second feathered descent.

With the feather deployed, *SpaceShipOne* dropped at a rate greater than 10,000 feet per minute [3,050 meters per minute]. However, it was extremely stable as it fell to the ground belly first.

"You could change the heading," Mike Melvill said. "If you were pointing at Cal City, you could turn around and point it to Mojave. And you used the elevons to do that. It was kind of weird because normally it would roll, but your sensation was that it was yawing."

"If you stepped on the rudders, it wasn't perceptible to you what was happening. Nothing happened. The only thing that really did anything was lateral spin. It was kind of neat to go over and look at a different view, and look over there and see what was over there. We did that a lot when we were flying as a glider in the atmosphere."

At 30,000 feet (9,140 meters) Melvill retracted and locked down the feather. *SpaceShipOne* was back as a glider, as shown in figure 7.6. He expanded the flight envelope for airspeed and g-force. And before landing, he executed *SpaceShipOne*'s first roll.

Departure from Controlled Flight (6G)

The focus of the test flight program now began to shift to prepare for the upcoming rocket-powered flights. Up to this point, *SpaceShipOne* was flown light, but for rocket-powered flight, it would have to maneuver with a fully fueled rocket engine. *SpaceShipOne* was loaded so the center of gravity (CG), or the single balance point of *SpaceShipOne*'s mass, moved to the aft to simulate these conditions.

When Melvill tested the stall characteristics for the aft-loaded *SpaceShipOne*, the nose swung upward uncontrollably before the wings reached the angle of attack at which they were expected to stall. *SpaceShipOne* entered into a spin while Melvill fought to regain control. Figure 7.7 shows *SpaceShipOne* as Melvill recovered from the tail stall.

"We had a pretty significant departure from controlled flight at high angle of attack, aft CG, due to a tail stall. That really was a big surprise," Doug Shane said.

Flight Test Log Excerpt for 6G

Date: 23 September 2003

	Flight Number	Pilot/Flight Engineer
SpaceShipOne	6G	Mike Melvill
White Knight	37L	Pete Siebold/Matt Stinemetze and Jeff Johnson

Objective: Third glide flight of *SpaceShipOne*. Aft CG flying qualities and performance evaluation of the spaceship in both the glide and reentry or "feather" mode. Glide envelope expansion to 95 percent airspeed, 100 percent alpha [angle of attack] and beta [sideslip angle], and 70 percent load factor. More aggressive post-stall maneuvering and spin control as a glider and while feathered. Nitrous temperature control during climb to altitude and performance of upgraded landing gear extension mechanism and space-worthy gear doors.

(source: Mojave Aerospace Ventures LLC, provided courtesy of Scaled Composites)

Fig. 7.6. After quickly correcting the avionics malfunction, *SpaceShipOne* and *White Knight* returned to the air several hours after the aborted fourth test flight. During this flight test, *SpaceShipOne* extended its feather for the first time. It performed superbly. *Mojave Aerospace Ventures LLC, photograph by David M. Moore*

The feather wasn't raised during the test flight, but during the climb to release altitude, the pressure test of the oxidizer tank revealed a variation of less than 6 psi. This meant that the temperature of the nitrous oxide inside could be controlled very well by exhaust air ducted in from *White Knight*.

Scaled Composites needed wind-tunnel data to evaluate the problem with the tail booms. "Except we didn't have a wind tunnel, but we did have a pickup truck. And we had our aero guy, Jim," Shane said.

Using a converted pickup truck fitted up with instrumentation, called the Land Shark, engineers aerodynamically tested mockups of the tail boom. With clearance from Mojave Airport, the Land Shark zoomed up and down a runway to collect data.

"We finally ended up doing a fence and a span increase on both the stabilizer and the elevon and resolved the problem," Shane said.

A triangular strake was also added to each tail boom, right in front of each horizontal stabilizer. *SpaceShipOne* was ready to go back to flight testing.

Back on Track (7G)

"A stall is when the air flowing over the wing no longer stays attached to the surface. It's not developing any lift anymore. And as soon as it stalls, you are not an airplane anymore. You are just a 2,000-pound [910-kilogram] lump falling out of the sky," Mike Melvill defined in *Black Sky*, the Discovery Channel documentary about *SpaceShipOne*.

For this flight, the only modifications to each tail boom were the additions of the strake and flow fence. The enlargement of the horizontal stabilizers would wait until the next test flight. However, the new modifications did improve the aerodynamics, and the uncommanded pitch-up of the nose at aft CG was eliminated. Melvill was able to then turn his attention to the feather and rocket-engine controls. Figure 7.8 shows the feather deployed as he continued to push maneuverability limitations.

After the functionality of the rocket-engine instruments and controls checked out, Melvill was ready to land. Figure 7.9 shows a view from the camera mounted in his helmet as *SpaceShipOne* neared the runway. The glide flight lasted 17 minutes and 49 seconds.

Flight Test Log Excerpt for 7G

Date: 17 October 2003

	Flight Number	Pilot/Flight Engineer
SpaceShipOne	7G	Mike Melvill
White Knight	38L	Pete Siebold/Cory Bird and David Moore

Objective: Fourth glide flight of *SpaceShipOne*. Primary purpose was to examine the effects of horizontal tail modifications at both forward and mid-range CG locations (obtained by dumping water from an aft ballast tank between test points). The tail modifications included a fixed strake bonded to the tail boom in front of the stabilator and a span-wise flow fence mounted on the leading edge of each stab at mid-span. Other test objectives included a functional check of the rocket motor controller, ARM, FIRE, and safing switches as well as the oxidizer dump valve. Additional planned maneuvers included full rudder pedal sideslips and more aggressive nose pointing while in the feathered configuration.

(source: Mojave Aerospace Ventures LLC, provided courtesy of Scaled Composites)

Fig. 7.9. Flying the first six piloted flights, two captive carries, and four glide flights, Mike Melvill continued to expand the flight envelope. Step by step, he pushed *SpaceShipOne* to perform a little harder so the engineers could get a more complete picture of its flying qualities. *Mojave Aerospace Ventures LLC, video capture provided courtesy of Discovery Channel and Vulcan Productions, Inc.*

A New Pilot Behind the Stick (8G)

Flying at 115 knots and 47,300 feet (14,420 meters), Pete Siebold detached from *White Knight*. Shown at the controls of *SpaceShipOne* in figure 7.10, Siebold became the second test pilot to fly *SpaceShipOne*. Mike Melvill had flown all the previous manned test flights.

Although *SpaceShipOne*'s performance was really starting to become dialed in, there was no shortage of work to do. Untested controls and the modified horizontal stabilizers needed evaluation—all this while Siebold learned how *SpaceShipOne* flew outside of the realm of the simulator. Figure 7.11 shows small strings attached to the horizontal stabilizer. With one free end, these strings helped the engineers identify how the air flowed over its surfaces.

Before Siebold completed the glide flight, he had to work on a new landing procedure that he and Binnie had been devising in order to improve reliability. "We had both very short, almost didn't make it to the runway, and very long, almost off the end of the runway, landing excursions," Doug Shane said.

Although *SpaceShipOne* glided okay, it didn't have anywhere near the performance or control of a sailplane effortlessly gliding upward on the tops of thermals. The procedure involved overflying the runway and hitting altitude waypoints while being able to accommodate coming into the approach too high or too low. After 19 minutes and 55 seconds in the air, *SpaceShipOne* landed at the intended aim-point.

Flight Test Log Excerpt for 8G

Date: 14 November 2003

	Flight Number	Pilot/Flight Engineer
SpaceShipOne	8G	Pete Siebold
White Knight	40L	Brian Binnie/Matt Stinemetze

Objective: The fifth glide flight of *SpaceShipOne*. New pilot checkout flight. Stability and control testing with the new extended horizontal tails. Tests included stall performance at aft limit CG and evaluation of the increased pitch and roll control authority. Other objectives included additional testing of the motor controller (MCS) and handling qualities in feathered flight.

(source: Mojave Aerospace Ventures LLC, provided courtesy of Scaled Composites)

Fig. 7.10. On November 14, 2003, Pete Siebold became the second test pilot to fly *SpaceShipOne*. The Scaled Composites team had to wear many hats. Siebold was also responsible for developing the software for the Tier One navigation unit (TONU) and flight simulator. *Mojave Aerospace Ventures LLC, video capture provided courtesy of Discovery Channel and Vulcan Productions, Inc.*

Fig. 7.11. During Pete Siebold's first flight, he had to evaluate additional modifications to the horizontal stabilizers put in place to rectify handling issues revealed two flights earlier. Small strings attached to the newly enlarged horizontal stabilizers were used to help analyze the air flow. *Mojave Aerospace Ventures LLC, video capture provided courtesy of Discovery Channel and Vulcan Productions, Inc.*

Fig. 7.12. Mike Melvill returned to the pilot's seat for the ninth flight, which was the sixth glide flight of *SpaceShipOne*. In the photograph, Melvill pulls a handle to his left to deploy the feather for additional evaluation. *Mojave Aerospace Ventures LLC, video capture provided courtesy of Discovery Channel and Vulcan Productions, Inc.*

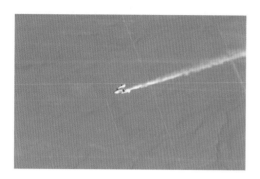

Fig. 7.13. An abort while fully fueled was a very big concern for Scaled Composites because of the heavy weight for landing. Even when testing *SpaceShipOne*'s performance with ballast to more closely match the weight of a fully fueled rocket engine, it was necessary to dump the ballast prior to landing. *Mojave Aerospace Ventures LLC, video capture provided courtesy of Discovery Channel and Vulcan Productions, Inc.*

Fig. 7.14. Among the biggest surprises to Scaled Composites was the difficulty of landing *SpaceShipOne* without undershooting or overshooting the runway. The test pilot had to manage the airspeed and altitude very carefully on approach to the runway for landing. *Mojave Aerospace Ventures LLC, video capture provided courtesy of Discovery Channel and Vulcan Productions, Inc.*

Flight Test Log Excerpt for 9G

Date: 19 November 2003

	Flight Number	Pilot/Flight Engineer
SpaceShipOne	9G	Mike Melvill
White Knight	41L	Brian Binnie/Cory Bird

Objective: The sixth glide flight of *SpaceShipOne*. Test pilot Mike Melvill's first flight with the enlarged tails. Emergency aft CG handling qualities eval and simulated landing exercise with the new tail configuration. Airspeed and g envelope expansion and dynamic feather evaluation.

(source: Mojave Aerospace Ventures LLC, provided courtesy of Scaled Composites)

Abort Contingencies (9G)

Melvill continued to expand the flight envelope and to test the feather. Figure 7.12 shows him inside the cockpit pulling the feather control. But he had another important task to complete.

All this flight testing was really aimed at one goal: *SpaceShipOne* flying a spaceflight. So, every step along the way had to accomplish something that would bring that goal just a little bit closer. But this also meant that contingencies had to be worked out. "Probably the biggest fear we had for every flight was having to abort," Doug Shane said about the rocket-powered flights.

Scaled Composites had very good confidence in how *SpaceShipOne* flew by this point. And after an extensive rocket-engine test program on the ground, they also had a good feeling about the rocket engine. So, Scaled Composites felt a safety incident was not likely to result from flying *SpaceShipOne* or firing the rocket engine. But an abort during a test flight was much more plausible, considering one had already occurred.

If *SpaceShipOne* aborted while full of fuel and oxidizer, then this would be much too much weight for it to handle during landing. "You

Fig. 7.15. The photograph shows Brian Binnie preparing right before his first time flying *SpaceShipOne*. This would be the last glide flight prior to the start of *SpaceShipOne*'s rocket-powered test flights. *Mojave Aerospace Ventures LLC, photograph by David M. Moore*

Fig. 7.16. Flight engineer Matt Stinemetze sits inside *White Knight* reviewing flight details as the time for liftoff nears. Stinemetze was the project engineer for *SpaceShipOne*, responsible for its assembly. *Mojave Aerospace Ventures LLC, photograph by David M. Moore*

have to actually dump nitrous to get rid of about 3,000 pounds (1,360 kilograms) of mass. But you couldn't deal with the rubber in the rocker motor," Shane said.

For this test flight, Melvill evaluated the emergency handling and landing characteristics. Figure 7.13 shows *SpaceShipOne* dumping the ballast, which it used to alter the CG during testing. Additional modifications were also made to the landing procedures, and figure 7.14 shows *SpaceShipOne*'s smooth touchdown.

Ready for Rocket Power (10G)

The rocket engine had been qualified only a few weeks previously and was ready to make its debut. Test flight 10G would be the last glide flight before attempting to light off the rocket engine with *SpaceShipOne*. Figure 7.15 shows Brian Binnie readying himself for his first flight in *SpaceShipOne*, and figure 7.16 shows flight engineer Matt Stinemetze making preparations in *White Knight* right before takeoff.

Flight Test Log Excerpt for 10G

Date: 4 December 2003

	Flight Number	Pilot/Flight Engineer
SpaceShipOne	10G	Brian Binnie
White Knight	42L	Pete Siebold/Matt Stinemetze

Objective: The seventh glide flight of *SpaceShipOne* and new pilot check out. Full functional check of the propulsion system by cold flowing nitrous oxide. Completed airspeed and positive and negative g-envelope expansion.

(source: Mojave Aerospace Ventures LLC, provided courtesy of Scaled Composites)

Fig. 7.17. *SpaceShipOne* was released at a height of 48,400 feet (14,750 meters). This video-capture image shows *SpaceShipOne* and *White Knight* mated up just prior to separation. The contrails from *White Knight's* turbojet engines can be seen trailing off in the background. *Mojave Aerospace Ventures LLC, video capture provided courtesy of Discovery Channel and Vulcan Productions, Inc.*

Fig. 7.18. Brian Binnie didn't have much time to enjoy the ride. As in all test flights, the pilots had a long list of tasks to complete, and after release, they only had 10–20 minutes of flight time. In preparation for the first rocket-engine test flight, Binnie performed a cold flow of the rocket engine, in which he followed all the procedures to run the rocket engine, except igniting the oxidizer and fuel. *Mojave Aerospace Ventures LLC, photograph by Scaled Composites*

Brian Binnie would now have a chance behind the control stick of *SpaceShipOne*. *White Knight* carried *SpaceShipOne* to the highest release altitude yet, 48,400 feet (14,750 meters). Figure 7.17 shows a close-up of *SpaceShipOne* attached to *White Knight* with the contrails from *White Knight*'s afterburners streamed away in the background.

A cold run of the rocket was performed. Binnie used all the controls and instruments for the rocket, including flowing the liquid N_2O oxidizer through the CTN (case/throat/nozzle) as if it were an actual rocket burn. However, without igniting the fuel, no combustion occurred.

After the successful cold run, Binnie turned his attention to completing the airspeed and g-force envelope expansion. Binnie pushed *SpaceShipOne* the hardest so far, and when finished, he glided back to Mojave, as shown in figure 7.18, eager for the next phase of flight testing to begin.

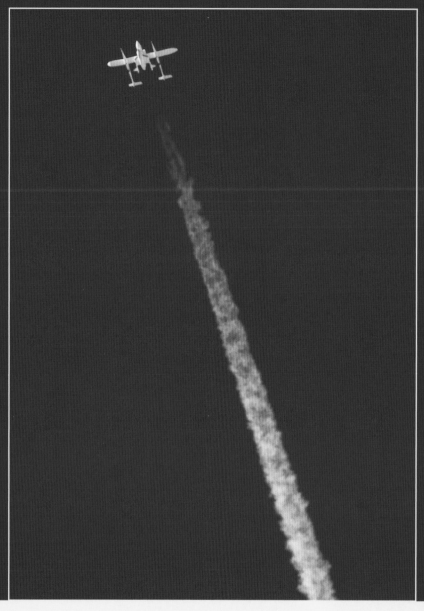

Up to this point, the contrails in the air only came from the twin turbojets on *White Knight*. After a total of ten unpowered flights, some captive carry and some glide, *SpaceShipOne* was ready to light its hybrid rocket engine for the first time in the sky. *Tyson V. Rininger*

Chapter 8

Rockets Ignite

Scaled Composites revealed *SpaceShipOne* to the public on April 18, 2003, but it had been running in secret for two years. On November 18, 2003, after about a year-long rocket-engine competition and seven glide flights, the hybrid rocket engine was qualified for flight testing in *SpaceShipOne*.

Flight testing is risky. Things go wrong as never-before-tested systems are evaluated or new conditions are encountered. It is all part of the process of shaking the design out. It is like an artist making a sculpture of an airplane from a block of wood. At first it doesn't look like anything, but slowly as pieces of wood are whittled away and form starts to take shape, little by little an airplane becomes more recognizable.

The process of flight testing is pretty much the same as testing other high-performance machines, whether it is a new racing sailboat or a new sports car. Nothing comes off the drawing board perfect. Flight testing is a flying laboratory. And as long as there are many more big steps moving forward than there are moving backward, good progress toward the goal is being made.

Scaled Composites attempted to become the first private company to go to space, yet the company had never even built an aircraft that broke the sound barrier. In fact, no private company had ever built an aircraft that broke the sound barrier as part of a non-governmental program.

The sound barrier, once thought impenetrable and its threshold guarded by demons, was smashed, along with the demons, by Chuck Yeager in the Bell X-1. Although this happened in 1947, the transition from subsonic to transonic to

Fig. 8.1. On December 17, 1903, Orville Wright, flying a twin pusher-propeller biplane with a canard, lifted off from the sandy dunes of Kitty Hawk, North Carolina. This was the first flight of a heavier-than-air, powered aircraft capable of sustained and controlled flight. *NASA*

supersonic was still a dangerous one. Before *SpaceShipOne* reached past the boundary of Earth's atmosphere, it would have to fly faster—several times faster—than the speed of sound.

"The powered flights were very different," Mike Melvill said. "As a glider, it is completely silent, and there is no noise at all. You don't have this overload of sensory audio. The audio side of your senses is really overloaded when the rocket motor starts. It's very, very noisy. It vibrates a lot. It really shakes you around in the airplane. It's a pretty exciting thing to do. The acceleration is enormous. You drop off the hooks and light the rocket motor. You get this enormous kick in your back. And right away, you turn the corner and point it straight up."

Flying above Mach 1 was not the same as flying below Mach 1. Flying outside the atmosphere was not the same as flying within the atmosphere. And flying with a rocket engine was not the same as flying without a rocket engine.

Flight testing would now have to be taken up a few notches.

Breaking the Sound Barrier (11P)

On December 17, a flimsy aircraft that no one thought would fly prepared to launch from sand-swept dunes. This wasn't the Mojave Desert, though. It was 1903 in Kitty Hawk, North Carolina. The Wright brothers proved that a couple of bicycle makers could accomplish the unbelievable—powered, controlled, and sustained flight. Their famous flight is shown in figure 8.1.

One hundred years later, to the day, Scaled Composites also looked to make the unbelievable believable. *SpaceShipOne* was ready for its first rocket-powered test flight.

The sound barrier posed the greatest challenge the team had yet to face. The choice of pilots was not a straightforward one. Mike Melvill, the most experienced test pilot with Rutan designs, had already piloted *SpaceShipOne* seven times. Pete Siebold and Brian Binnie had each only flown *SpaceShipOne* once. During such a critical juncture, the test pilot with the greatest stick time in *SpaceShipOne* would seem like the most likely candidate. But Binnie was selected. The entire fate of Rutan's space program now rested in his hands.

Flight Test Log Excerpt for 11P

Date: 17 December 2003

	Flight Number	Pilot/Flight Engineer
SpaceShipOne	11P	Brian Binnie
White Knight	43L	Pete Siebold/Cory Bird

Objective: The eighth flight of *SpaceShipOne* and first powered flight. fifteen-second burn of the rocket motor and supersonic flight. Motor light-off at altitude and inflight engine performance. Vehicle handling qualities through transonics and feather performance from altitude.

(source: Mojave Aerospace Ventures LLC, provided courtesy of Scaled Composites)

One of the most obvious reasons for the selection was Binnie's experience flying supersonic jets for the U.S. Navy. "I would suspect that certainly was part of the thinking that went into it," Binnie said. "I was also kind of the anointed test director for the rocket motor. So, I was very familiar with all the ins and outs with the motor and its performance. I was comfortable with its design and robustness and performance, more so than perhaps anybody else. And perhaps I had, right from the first time I saw it fire, this image in my mind of how it would play out when put into a vehicle. I was very motivated to fly that flight."

Binnie also had flown *Roton*, a spaceship prototype produced by Rotary Rocket, Inc., that was half rocket and half helicopter. "That vehicle was decidedly scary to fly," Binnie said. "I flew it three times. I thought, 'You know, if I have to do this one or two more times, I'm pretty sure I'm going to kill myself.' I was honestly sort of fearful that I'd climb into this thing and not climb out. I never felt that way with the spaceship. There were certainly things that could go wrong that could cause an accident, but I never felt that the accident was going to be life threatening."

With Pete Siebold flying *White Knight*, Cory Bird released *SpaceShipOne* at 47,900 feet (14,600 meters). Dropping down, Binnie stabilized *SpaceShipOne* and started to pull the nose up.

"We knew very little about it. It was all about risk. Our standard for success was, literally, if the rocket motor lit off even for just a few seconds and we shut it down, we were going to call that a success," Binnie said.

How would the rocket engine function? Would *SpaceShipOne* hold together during the enormous acceleration? What about the messy transonic region? How about the cockpit? Would it shake too much or be too loud to operate the vehicle? How would the untested controls work at supersonic? Would *SpaceShipOne* climb quickly enough so that an overspeed didn't rip it to pieces?

"This was the riskiest flight of our whole program, probably times five or ten," Doug Shane said.

At 44,400 feet (13,530 meters) and flying Mach 0.55, Binnie lit the rocket engine. "It's a real rush to kind of ride something that powerful. Like I said, it is like a tsunami comes through the cabin. You are just taken away. So, it is very exhilarating," Binnie said.

As *SpaceShipOne* rocketed upward, it broke the sound barrier after only 9 seconds. Figure 8.2 shows *SpaceShipOne*'s rocket engine firing.

Binnie shut down the rocket engine after a burn duration of 15 seconds as *SpaceShipOne* hit a maximum speed of Mach 1.2 (800 miles per hour or 1,290 kilometer per hour) while having pulled more than 3 g's. As *SpaceShipOne* became the first manned vehicle to have ever used a hybrid rocket engine, it pointed up at 60 degrees and still climbed. *SpaceShipOne* continued to coast upward and was upside down by the time it reached near-weightlessness at an apogee of 67,800 feet (20,670 meters). Binnie then descended for about a minute with the feather deployed and retracted it at 35,000 feet (10,670 meters).

"It was manageable," Binnie said. "It wasn't the most graceful flight. If you looked at it from inside the cockpit, there were a lot of snakes being killed. The boost part of it went extremely well, all things considered, and gave us a lot of confidence that we had a design that was capable. It was only the landing that marred an otherwise wonderful day."

Fig. 8.2. Brian Binnie took the controls of *SpaceShipOne* for test flight number 11P, one hundred years after the flight of the first successful airplane, in which the 1903 Wright Flyer had flown for 12 seconds. Even though Binnie burned the rocket engine for only 15 seconds, *SpaceShipOne* broke the sound barrier after only 9 seconds. *Mojave Aerospace Ventures LLC, video capture provided courtesy of Discovery Channel and Vulcan Productions, Inc.*

Fig. 8.3. A crash landing blemished an otherwise successful first rocket-powered test flight. *SpaceShipOne* touched down harder than the landing gear could handle, causing the left rear main landing gear to collapse and *SpaceShipOne* to skid off the runway. *Mojave Aerospace Ventures LLC, video capture provided courtesy of Discovery Channel and Vulcan Productions, Inc.*

On the landing, which had proved to be a surprising and persistent source of trouble for Scaled Composites, *SpaceShipOne* hit very hard. The left main landing gear collapsed, and *SpaceShipOne* skidded to the left and off the runway into the dirt, as shown in figure 8.3 and figure 8.4.

Burt Rutan gave some clarification about the cause of the crash landing during a press conference late in 2004. He said, "We were testing at that time, a flight control modification that would give us a better margin on flutter in the transonic area. And that damper in the flight controls had gotten so cold that the controls were extremely sticky. That landing wasn't Brian's fault. It was the fault of us not doing the proper thing, and that is putting a heater on the damper so that it didn't freeze up on him on the landing approach."

Fig. 8.4. Unknown to Brian Binnie at the time, a modification to part of the mechanical flight control system malfunctioned just before landing. *SpaceShipOne* felt like it was getting ready to stall, so Binnie was put into a seemingly no-win situation. If *SpaceShipOne* did in fact stall, then there wasn't enough altitude to recover. But to prevent a stall, he'd have to put the nose down to pick up speed. *Mojave Aerospace Ventures LLC, photograph by David M. Moore*

As Binnie was flaring up for the landing, the damper unfroze. The controls buffeted like *SpaceShipOne* was going to stall. It was actually just the control linkages responding differently, but there was no way that Binnie could have known this. So, he dropped the nose to prevent the stall that he thought was about to occur. As a result, *SpaceShipOne* came in too fast and too hard for the fragile landing gear.

Unplugged (12G)

After a little less than three months, *SpaceShipOne* was ready to return to the air again, with Pete Siebold at the controls. However, its rocket engine would be quiet during this test flight.

"It was after the famous 11P flight, which resulted in significant damage to the aircraft on the hard landing," Siebold recalled. "It was in one respect what we would call a functional check flight after any major modifications to the airplane. We wanted to go fly it in a semi-benign environment and try and shake down any of the problems that we may have overlooked or additional problems that had been created due to the modifications.

"The other reason was we made some modifications to the aero-dynamic shape. We added the thermal protection to the aircraft. If you look at the artwork of that flight, it shows the red leading edges and shows the TPS addition. That actually changed the wing shape slightly and the aerodynamic shape. So, we wanted to go and fly that and see if there were any ill effects to that modification for the flight."

Siebold started off his second time flying *SpaceShipOne* at an altitude of 48,500 feet (14,780 meters), which was the highest point that *White Knight* ever released *SpaceShipOne*. Scaled Composites had originally planned on releasing *SpaceShipOne* from an altitude of 50,000 feet (15,240 meters). However, *White Knight* had a very difficult time flying this high and too frequently had come out of afterburners or flamed out altogether at an altitude even below this one. The lower launch altitude did actually work in *SpaceShipOne*'s favor.

During powered flight, the first thing that *SpaceShipOne* had to do was "turn the corner" as soon as it possibly could, but the higher the altitude, the less air there was for the wings to bite into in order to make a quick turn upward. So, in terms of utilizing the energy from the rocket engine as efficiently as possible, launching below 48,000 (14,630 meters) feet gave better overall performance, since *SpaceShipOne* would spend more time pointing up than over.

However, since Siebold wasn't concerned with lighting off the rocket, he wanted all the altitude he could get for the glide flight. "There were some minor glitches," he said. "The thermal protection system started cracking at low temperatures, and I think there was actually a formulation change made between that flight and the actual powered flight."

But the thermal protection system (TPS) wasn't the only system being checked out. Siebold also evaluated the reaction control system (RCS) that would be used to maneuver *SpaceShipOne* while in the absence of the atmosphere. This and other testing worked out smoothly, and *SpaceShipOne* touched down safely, even in the presence of a strong crosswind.

A Third of the Way There (13P)

Why did the desert tortoise cross the runway? Now that Scaled Composites planned a longer burn, it could no longer fly under the radar of the FAA. Scaled Composites needed a commercial launch license, and the Office of Commercial Space Transportation (AST) of the FAA required Scaled Composites to do an environmental impact report as part of the application process.

Doug Shane explained, "One of the caveats that came back with that [application] is before taking off in *White Knight* and before landing in *SpaceShipOne*, we had to do a sweep of the Mojave runway for desert tortoises. And if we found a desert tortoise, we couldn't move it. We couldn't touch it. We couldn't talk to it. We couldn't negotiate with it. We couldn't threaten it. We couldn't bribe it in any way.

"What we had to do was call the desert tortoise control specialist from Ventura County, about a three-hour drive away, and let them come and negotiate some kind of a successful conclusion."

Fortunately, Scaled Composites got their license, no tortoises made runway excursions, and flight 13P was a go. Pete Siebold, who would have the controls of *SpaceShipOne* for the second power flight, recalls, "That was the first time that we flew basically at our heavy weight. First time we put all the nitrous on board. The airplane was fully ballasted to be a representative weight for the spaceflights. It was part of that incremental weight expansion."

Right upon release from *White Knight* at 45,600 feet (13,900 meters) and 125 knots, though, *SpaceShipOne* ran into problems. "We pulled the nose up to maintain our speed, and we realized that the wings at that weight and speed could not lift the vehicle," Siebold said. "So, the wings were stalling earlier than anticipated. So, there was this problem that we were faster than we wanted to be to light the rocket, which would result in an overspeed. But we also didn't want to abort the flight, because we had some really questionable handling qualities if we dumped all the nitrous to our landing weight. It would send our CG dangerously far aft."

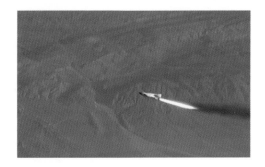

Fig. 8.5. Nearly four months after the first rocket-powered flight test, Pete Siebold ignited the rocket engine of *SpaceShipOne* for 40 seconds and reached an apogee of 105,000 feet (32,000 meters). He hit a top speed of Mach 1.6 on the boost and Mach 0.9 on the way down. *Mojave Aerospace Ventures LLC, video capture provided courtesy of Discovery Channel and Vulcan Productions, Inc.*

Fig. 8.6. *SpaceShipOne* reached a third of the way up to the Ansari X Prize goal of 328,000 feet (100,000 meters). This was all part of the incremental testing plan. Although not able to test the feather while moving supersonically, Pete Siebold was high enough to test the reaction control system (RCS). *Mojave Aerospace Ventures LLC, video capture provided courtesy of Discovery Channel and Vulcan Productions, Inc.*

Ignition was delayed by two minutes as Mission Control tried to decide whether to abort the flight or just to go ahead with the burn.

"One of the benefits we had in delaying was we went to a lower altitude, which would allow us to turn the corner much faster, which minimized the risk of overspeed in starting the flight at a higher than expected airspeed," Siebold said.

After dropping for more than a mile, Siebold lit the rocket engine at 38,300 feet (11,670 meters). Figure 8.5 shows Siebold during the pull-up after rocket engine ignition.

SpaceShipOne was moving at Mach 1.6 when Siebold shut down the rocket engine after a 40-second rocket burn. The rocket plane reached an apogee of 105,000 feet (32,000 meters), which was about a third of the distance *SpaceShipOne* needed to climb for the Ansari X Prize. On descent, *SpaceShipOne* experienced Mach 0.9 while feathered.

The flight overall was a success. Burn duration increased significantly from the 15 seconds of the first rocket-powered flight.

"We also were able to demonstrate that we could maintain control during the pull-up, which was something on 11P that was sort of in question. Brian was fighting the vehicle trying to keep the wings level. So, overall, I think the only objective that we weren't able to meet was the supersonic feather reentry" Siebold said.

Figure 8.6 shows Siebold flying *SpaceShipOne* back to Mojave.

Feather at Supersonic (14P)

Nearly one full year since flight testing began, *SpaceShipOne* was on the verge of making a spaceflight. One critical piece of information was missing, though. "The object of that flight was to do a supersonic feathered reentry," Mike Melvill said. "We needed that data before we could go beyond that." Figure 8.7 shows *SpaceShipOne* mated up to *White Knight* in preparation for the third powered flight.

Ten seconds after releasing from *White Knight* at 46,000 feet (14,020 meters), Melvill lit off the rocket engine. Figure 8.8 shows a dramatic rearward view of the rocket engine's fiery plume and exhaust.

"During the boost after he reached the vertical part of the trajectory, the avionics display started flickering and then went blank," Doug Shane said. "We all had good displays in the ground station. And Mike said, 'I looked out the window, and we were going pretty much straight up. So, I stayed with her.' Gotta love a guy like Mike. Of course it came back on as soon as the motor shut down."

The rocket engine burn duration was set by a timer. As Melvill looked out the windows to navigate, *SpaceShipOne* boosted to 150,000 feet (45,720 meters) and Mach 2.5, and then its rocket engine shut down. *SpaceShipOne*

Flight Test Log Excerpt for 14P

Date: 13 May 2004

	Flight Number	Pilot/Flight Engineer
SpaceShipOne	14P	Mike Melvill
White Knight	56L	Brian Binnie/Matt Stinemetze

Objective: The third powered flight of *SpaceShipOne*. 55 seconds motor burn time. Handling qualities during boost and performance verification. Reaction control system use for reorientation to entry attitude. Supersonic feather stability and control.

(source: Mojave Aerospace Ventures LLC, provided courtesy of Scaled Composites)

Fig. 8.7. The photograph shows *SpaceShipOne* being prepared for its third rocket-powered test flight. *SpaceShipOne* had made its very first test flight just about a year earlier. The main goal of this test flight was to evaluate the performance of the feather at supersonic speed. *Mojave Aerospace Ventures LLC, photograph by David M. Moore*

Fig. 8.8. *SpaceShipOne* had several video cameras mounted inside the cockpit and on its exterior. This video-capture image shows a rearward view from the top of the fuselage of the rocket engine's plume and exhaust. *Mojave Aerospace Ventures LLC, video capture provided courtesy of Discovery Channel and Vulcan Productions, Inc.*

continued its ascent to an apogee of 211,400 feet (64,430 meters). But since Melvill had lost his avionics during boost, the trajectory was not exactly spot on. "I was doing forward loops, or something, at the top. It slowed down but came back in, and then it was swinging around a lot."

Melvill used the RCS to dampen the oscillations. In the feather configuration, *SpaceShipOne* reentered the atmosphere at Mach 1.9 and 3.5 g. *SpaceShipOne* quickly stabilized and made its feathered "carefree"

reentry just as expected. It actually descended more smoothly at supersonic speeds than it did at subsonic speeds. Figure 8.9 shows *SpaceShipOne* above Earth, with Los Angeles and the California coastline in the background.

Back on the ground, engineers traced the avionics malfunction to a dimmer, a small electrical component. And since the thermal protection data looked good, Scaled Composites felt that *SpaceShipOne* performed well enough to continue forward.

First Commercial Astronaut (15P)

At 6:47 a.m. PST on June 21, 2004, *White Knight* lifted off from Mojave's Runway 30, as shown in figure 8.10, to the cheers of twelve thousand spectators. Several days earlier, Mojave Airport officially became Mojave Air and Space Port after receiving its launch site operator license. It would have been another typical windy desert early morning with the Sun still not fully awake, except for a gangly looking airplane toting a stout little rocket plane on the first part of its journey to the hopeful reaches of space. Figure 8.11 shows the mated pair spiraling up to the launch altitude.

At an altitude of 47,000 feet (14,330 meters), *White Knight* could no longer guide *SpaceShipOne* any further and cut the rocket craft free to continue the quest on its own.

"You release the back pressure, and then the airplane starts to climb," Mike Melvill said. "And at that point you must have the motor running. You unguard the switch that turns the electrons on to the ignition system, and you unguard the switch that fires it.

"A rocket motor doesn't start off like a jet engine. It starts off as hard as it is ever going

to be right there. The first time it lights, it is going as strong as it will ever be. In fact, it gets weaker as you go along. The initial kick on your back is very strong—more than 3 g's. Your eyeballs go in at 3 g's. And then you make about a 4-g turn. And you do that by just pulling back on the stick. In only 9 or 10 seconds, you are supersonic. You are supersonic about two-thirds of the way through that turn. And then you can't use the stick anymore. You have to use the trim. So, that transition is something you have to learn in the simulator."

But before completion of the pull-up, severe wind shear rolled *SpaceShipOne* to the left. As Melvill tried to regain wings level, *SpaceShipOne* rolled 95 degrees to the right and then 90 degrees to the left. He soon got control and had *SpaceShipOne* flying vertical, as in figure 8.12. But he ended up far off course.

A minute into the burn, as the oxidizer ran low in the tank, it began to transition from a liquid to a gas. The roar of the rocket engine made a drastic change. "As it starts sucking gas, it chugs," Melvill explained. "And it goes boom, boom, boom, boom, boom, boom, boom, boom, boom, like that as it is going up. It really rattles your head. It doesn't do that for too long. But it is disconcerting the first time it happens. I didn't know what the heck that was all about."

But the rocket engine wasn't finished with its mischief. "The rubber that is on the inside of that thing has got ports in it, pie-shaped ports that run the whole length of the rubber fuel. So, as you are burning all the surfaces of the inside of each of these pie-shaped ports, they are coming together. You end up with a plus-sign–shaped piece of rubber that is very thin that runs the whole length of the machine. It will break off eventually and go out the back. That must have broken off and got sideways in the nozzle or something because it made a tremendous bang. It really rattled the airplane. I thought the whole tail had fallen off the airplane."

To improve the airflow around the rocket engine for this flight, a fairing was added, which extended from the back of the fuselage over the sides of the nozzle. But heat from the rocket engine caused it to buckle. It was necessary to modify the design slightly for the next flight. It was unlikely the source of the sound Melvill heard.

Melvill seriously wondered if he'd be able to make it back, but Mission Control could see that the tail booms were just fine from the live video feed of the onboard camera. Unfortunately, Melvill didn't have the capability to view this imagery. A failure of the primary pitch trim control, which Melvill needed while moving at supersonic speed, also occurred while the rocket engine blasted away. This only

Flight Test Log Excerpt for 15P

Date: 21 June 2004

	Flight Number	Pilot/Flight Engineer
SpaceShipOne	15P	Mike Melvill
White Knight	60L	Brian Binnie/Matt Stinemetze

Objective: First commercial astronaut flight by exceeding 100 kilometers (328,000 ft).

(source: Mojave Aerospace Ventures LLC, provided courtesy of Scaled Composites)

Melvill

15P   328 000   60L
71

Binnie/Stinemetze
2.9/328

Fig. 8.10. As *White Knight* fired up its engines and began taxiing to the runway, it rounded the corner, around the control tower, to find a crowd of twelve thousand well-wishers lined up along the flightline. At 6:47 a.m. on June 21, 2004, *SpaceShipOne* began its first journey into space. *Tyson V. Rininger*

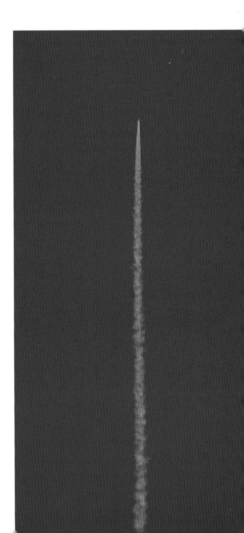

Fig. 8.12. Mike Melvill ignited the rocket engine, "turned the corner," and blasted nearly straight up. The rocket engine burned for 76 seconds and shut down at 180,000 feet (54,860 meters). After reaching a maximum speed of Mach 2.9, *SpaceShipOne* coasted the rest of the way up to space. *Tyson V. Rininger*

Fig. 8.11. With Brian Binnie flying *White Knight*, the captive carry lasted about one hour before the separation altitude of 47,000 feet (14,330 meters) was reached. But for Mike Melvill, sitting in the cockpit of *SpaceShipOne*, the hour seemed much, much longer. *Tyson V. Rininger*

Fig. 8.13. *SpaceShipOne* reached an apogee of 328,491 feet (100,124 meters), barely above the Ansari X Prize goal. During Mike Melvill's 3.5 minutes of weightlessness, he released two handfuls of M&Ms into the cockpit to float around in zero gravity. *Mojave Aerospace Ventures LLC, video capture provided courtesy of Discovery Channel and Vulcan Productions, Inc.*

deepened his worry. However, he was able to get pitch trim control back by switching to the backup.

Melvill obviously had his hands full, and he wasn't finished battling the rocket engine. He said, "The thrust is supposed to be right down the centerline. The nozzle is an ablative nozzle. It would burn away, and if it burns away a little bit on one side compared to the other side, you can end up with a yaw going on or a pitch. It burned like about a half a degree from being straight.

"Full rudder would only just deal with a half a degree from asymmetry in the thrust. I put the rudder on the floor, but it still went sideways across the ground. It is just amazing how quickly you go from one place to another if that happens."

By this point, the good-luck horseshoe pin that Sally Melvill pinned to her husband's flight suit just before the flight must have weighed five pounds.

"Poor trajectory control drew the airplane way far downrange. It spent an awful lot of energy going downrange rather than straight up where you want it to go," Doug Shane said.

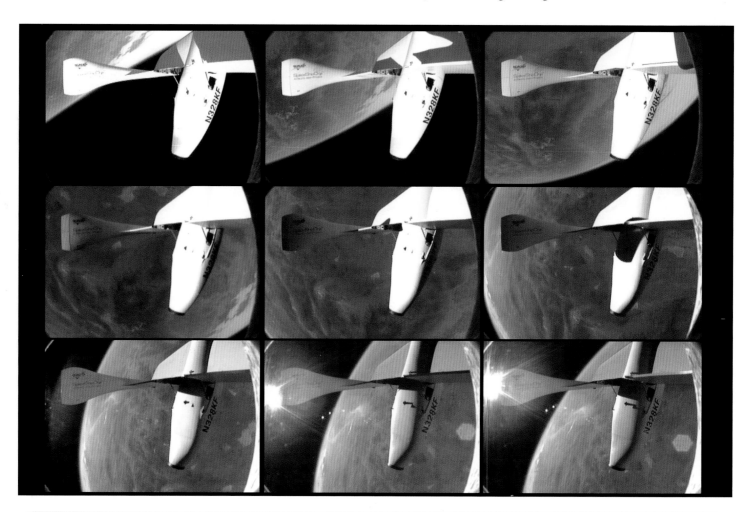

Fig. 8.14. This sequence of images, taken at three-second intervals, shows *SpaceShipOne* zooming over Earth during a period of less than thirty seconds. The shadow of the feather on *SpaceShipOne* quickly moved position as *SpaceShipOne*'s orientation to the Sun rapidly changed. *Mojave Aerospace Ventures LLC, screen captures provided courtesy of Discovery Channel and Vulcan Productions, Inc.*

Boosting *SpaceShipOne* to Mach 2.9 (2,150 mile per hour or 3,460 kilometers per hour), the rocket engine cut off at 180,000 feet (54,860 meters) after firing for 76 seconds. *SpaceShipOne* coasted toward space, but concern rapidly grew whether or not it would reach the Ansari X Prize goal of 328,000 feet (100,000 meters).

Paul Allen could only watch and wait as the Mission Control staff tracked *SpaceShipOne*'s progress. "You've got a human being in a very small projectile that is going straight up at Mach 3," Allen said, "and me pacing around behind the scenes going, 'I just want Mike to get back on the ground safely.' And you got an altimeter that just zips around as *SpaceShipOne* accelerates upward during the engine burn. And then it starts coasting. The altimeter is just wrapping itself around itself as fast as it can, and then it starts slowing. You are wondering if *SpaceShipOne* is going to get high enough, and then it did but barely."

Before reaching apogee, Melvill raised the feather in preparation for reentry. Gravity was calling *SpaceShipOne* back, and the spacecraft crept up slower and slower until it finally stopped. Measurements of the altitude were being taken from several different sources, and for a moment, there was nothing but uncertainty. *SpaceShipOne* did beat the mark by just a fraction, making it to 328,491 feet (100,124 meters), but substantially short of its targeted 360,000 feet (109,700 meters).

Despite the problems Melvill faced in the cockpit, he had a few moments to enjoy the 3.5 minutes of weightlessness. While in zero-g, he wanted to give a good demonstration of the weightless

Fig. 8.15. After battling wind sheer, an ornery rocket engine, a buckled engine fairing, and a primary pitch-trim malfunction, Mike Melvill had an uneventful reentry. But since *SpaceShipOne* was off course going up, it reentered about 30 miles (48 kilometers) south of Mojave. This was well within *SpaceShipOne*'s glide range, though. *Tyson V. Rininger*

Fig. 8.16. Burt Rutan welcomed his longtime friend Mike Melvill back from space with a great big hug. Paul Allen didn't hold back either. With the Scaled Composites team and the backing of Allen, Rutan proved he could design a spaceship and get it safely into space. *Tyson V. Rininger*

experience. From a zipped pocket on the left arm of his flight suit, Melvill grabbed two handfuls of M&Ms, which he had bought on the way to the airport earlier that morning, and cast the multicolored candies out into the cockpit. Figure 8.13 shows Melvill in the cockpit with M&Ms floating about.

"The sky was jet black above, and it gets very light blue along the horizon. And the Earth is so beautiful, the colors of the Earth, the colors of the high desert, and along the coastline. And all that fog or low stratus that's over L.A. looked exactly like snow. The glinting and the gleaming of the Sun on that low cloud looked to me exactly like snow," recounted Melvill at a press conference after the spaceflight.

"And it was really an awesome sight. I mean, it was like nothing I've ever seen before. And it blew me away. It really did."

Figure 8.14 shows video frames, at three-second intervals, of *SpaceShipOne*, with its feather extended, as it races through space over

Earth. In just a matter of seconds, it moves from one side of Earth to the other. Notice how the shadows change position as the orientation of *SpaceShipOne* rapidly changes with respect to the Sun. By the last frame, the Sun is behind the portside tail boom.

With the pitch trim control anomaly resolved, all Melvill had to do was let *SpaceShipOne*'s feather handle the "carefree" return into the atmosphere. *SpaceShipOne* hit 5.0 g's while reaching Mach 2.9 during reentry.

"We started out over Boron and wound up directly over the top of the Palmdale VOR [VHF omnidirectional radio beacon]. That is a long way south, right out of the restricted area," Melvill said. In fact, *SpaceShipOne* reentered over Palmdale Airport at 65,000 feet (19,810 meters), some 30 miles (48 kilometers) south of Mojave Air and Space Port.

"It was perfectly safe to be flying as an airplane or glider out there," Doug Shane said.

Fig. 8.17. In a surprise presentation, Patti Grace Smith, the FAA's associate administrator for Commercial Space Transportation, awarded Mike Melvill the first-ever commercial astronaut wings. In flying *SpaceShipOne* above 328,000 feet (100,000 meters), Melvill satisfied the primary Tier One goal of getting to space. Now it was time to set the sights on the prize. *Mojave Aerospace Ventures LLC, photograph by David M. Moore*

SpaceShipOne had better than a 60-mile (97-kilometer) glide range. It defeathered at 57,000 feet (17,370 meters) and started to glide back to Mojave.

"I got back to Mojave at 40,000 to 50,000 feet [12,190 to 15,240 meters]. I think we could make it from L.A., LAX probably, if we ended up really off course," Melvill said.

Doug Shane would eventually get a call from the FAA for a meeting. Palmdale was the location of the air traffic control center responsible for all of Southern California. When Shane met with the FAA, an official enthusiastically said Scaled Composites could have all of LA Center's airspace, no questions asked. All it would cost is one ride.

Figure 8.15 shows *SpaceShipOne* coming in for a perfect landing after a tumultuous spaceflight. As the spacecraft came to rest, completing only its fourth powered flight, Melvill, who had worked for Burt Rutan since 1978 and been Rutan's first employee, became the first commercial pilot of a vehicle to and from space. The Scaled Composites team became the first nongovernment space program to successfully send a human to space. And figure 8.16 shows Burt Rutan and Paul Allen congratulating Melvill on accomplishing a true milestone of flight. Also waiting to congratulate Melvill was Apollo astronaut Buzz Aldrin, who welcomed him to the club. Melvill became only the 433rd person in space since Cosmonaut Yuri Gagarin was the first to reach space in 1961. This works out to an average of ten new people to reach space per year since the very start of human spaceflight.

During a presentation shortly after the spaceflight, the FAA had a surprise for Melvill. "I am very pleased and honored to present, for the very first time, these FAA commercial astronaut wings to Mike Melvill in recognition of this tremendous achievement," said Patricia "Patti" Grace Smith, associate administrator for the Office of Commercial Space Transportation.

Figure 8.17 shows Melvill with his astronaut wings standing next to Paul Allen and Patti Grace Smith. "I wasn't expecting anything like that," said Melvill. "It was really a thrill."

The FAA now issues astronaut wings to any member of a crew, including passengers, on a spacecraft that exceeds an altitude of 50 miles (80.5 kilometers) during a spaceflight. The 50 miles (80.5 kilometers) was an arbitrary boundary traditionally used by the U.S. military. Internationally, the bar is higher, and the accepted boundary of space, which was used for the Ansari X Prize, is set by the scientifically based Karman Line of 100 kilometers (62.1 miles).

The primary goal of the Tier One space program set by Burt Rutan and Paul Allen had been achieved once *SpaceShipOne* returned safely to Mojave after reaching space. To that end, *SpaceShipOne* was optimized for altitude, not payload. Any unnecessary weight would have adversely impacted performance. So, even though *SpaceShipOne* made it past the Ansari X Prize altitude goal, the passenger requirement wasn't met.

Scaled Composites had much to think about now that the focus of Tier One was ready to change. Their first spaceflight had not gone entirely as planned, but the Ansari X Prize was now legitimately within reach. Yes, *SpaceShipOne* had made it to the altitude required by the Ansari X Prize—with only a few hundred feet to spare. Serious problems were encountered, however, and *SpaceShipOne* wasn't even hauling the extra weight of a payload. It was necessary to review the spaceflight data and evaluate and repair the damage to *SpaceShipOne*. Corrective action was required to ensure these mishaps would not repeat and that the performance was improved to meet the demands of the Ansari X Prize.

The Ansari X Prize was to expire in half a year. Within that time, Scaled Composites had to fly two qualifying flights. Six months sounds like a lot of time, but margins of error for spaceflight are razor thin. For the challenges they faced—risking a test pilot's life and possibly derailing the drive for private spaceflight for many years if something catastrophic occurred—six months were more like six weeks. The need for an additional powered flight before making an attempt at the Ansari X Prize had to be considered.

Brian Binnie and Mike Melvill each had flown *SpaceShipOne* on famous firsts. Binnie was first to fly a rocket-powered flight and to break the sound barrier. Melvill was first to fly captive-carry and glide flights. He also took *SpaceShipOne* to space. But Pete Siebold had flown *SpaceShipOne* one more time than Binnie and was also more current. Who would fly for the Ansari X Prize? *Mojave Aerospace Ventures LLC, photograph by Scaled Composites*

Chapter 9

Capturing the Anasari X Prize

Time was running out. Scaled Composites announced that they would be making their attempt at the Ansari X Prize on September 29, 2004. To win, *SpaceShipOne* would have to fly not just on that day, but would have to fly once more by October 13th, leaving less than three months before the Ansari X Prize would expire. A major setback would take *SpaceShipOne* out of contention.

Brian Feeney's team scheduled to launch their spacecraft *Wild Fire* on October 2, 2004. But there was a concern as to whether or not they would be ready to launch. What would happen if the January 1, 2005, expiration rolled by without a winner?

"We had a contingency plan that whoever won it would get a trophy but not ten million bucks," Erik Lindbergh said. "But whether or not it would have been as effective was a question. Whether or not it would have gained as much media attention was a question. And also whether or not we would have been able to keep the doors open was a question."

The X Prize Foundation wanted desperately to award the Ansari X Prize. To them, this wasn't a one-shot deal. Winning the Ansari X Prize meant jumping the first, but highest by far, hurdle on the path to public space access. But having the prize unclaimed was not their only fear. They knew that progress would only come from the successes and the failures of flying over and over again.

"It was very tense the night before and in the morning as we were gathering in the cold in Mojave to watch the launch," Anousheh Ansari said. "We were very anxious. We had to prepare ourselves for all sorts of possibilities."

Fig. 9.1. When Mike Melvill made it to space in *SpaceShipOne*, it was like a great awakening. Scaled Composites gave proof to the world that commercial spaceflight was for real. Millions of people seemed to catch the space craziness as the Ansari X Prize attempts were made and broadcasted live on television and over the Web. *Dan Linehan*

There was no doubt of the risks involved with spaceflight. And although *SpaceShipOne*'s first spaceflight earlier that June had some unexpected difficulties, the spacecraft and the pilot made it back safely. This was no guarantee for subsequent missions, which would continually stretch the flight envelope. After all, *SpaceShipOne* was still a research vehicle.

Everyone felt the enormity of the events. "To watch how the wives said goodbye to their husbands before they went up and wished them well was certainly a moment when you felt the responsibility of being involved in a project like this," Paul Allen said, "and them being worried for their husbands and you being worried, too."

The attempts at the Ansari X Prize would have unprecedented media coverage as well. Tens of thousands of people gathered to watch the launches in person (refer to figure 9.1 and figure 9.2). And the launches were broadcast live over television and the World Wide Web in a way like none other.

"The whole world was watching," said Gregg Maryniak, the executive director of the X Prize Foundation. "Most people don't appreciate that this was the first spaceflight ever that had video coming down that people—regular people—were watching in real time from a manned spaceship. It has never happened before. It has happened where in flight you could see snippets but never during ascent. When was the last time you saw NASA showing footage from inside the shuttle?"

The Ansari X Prize received upwards of six billion media impressions over its course, with a large percentage of them focused

on the small company from the Mojave Desert that was ready to prove that its first spaceflight wasn't a novelty.

"The entire Tier One team that was taking this vehicle through flight testing had been working really, really hard for the last six weeks or so to where there was almost always someone at Scaled doing something with the vehicle or preparing for the flights or in the simulator," Brain Binnie said.

"We had a lot of anxiety between our first flight to space with Mike with the lightweight vehicle and trying to decide how we were going to make the adjustment to carrying 600 pounds [270 kilograms] of pay-load for the X Prize flights and still get to those same altitudes. There was concern that the

Flight Test Log Excerpt for 16P

Date: 29 September 2004

	Flight Number	Pilot/Flight Engineer
SpaceShipOne	16P	Mike Melvill
White Knight	65L	Brian Binnie/Matt Stinemetze

Objective: First X Prize flight: ballasted to simulate 3 place and to exceed 100 kilometers (328,000 ft).

(source: Mojave Aerospace Ventures LLC, provided courtesy of Scaled Composites)

Melvill – X₁
16P   65L
337.4/3.0

rocket motor didn't have enough energy or impulse for us to get there. We had spent a lot of time worrying about that, wondering whether we needed to augment the motor with some other boosters. We eventually settled on a scheme that was really quite clever, but it took a while to work out the details."

X1: The First Ansari X Prize Flight (16P)

"Ladies and gentlemen, we are at the start of the personal spaceflight revolution, right here, right now. It begins in Mojave, today. What is happening here in Mojave today is not about technology. It is about a willingness to take risk, to dream, and possibly to fail," said Peter Diamandis during the morning of September 29, 2004, as X1, the name of the first required flight of *SpaceShipOne* in the quest for the Ansari X Prize, prepared to launch.

Mojave was abuzz. A little more than three months earlier, Mike Melvill had earned his astronaut wings as he piloted *SpaceShipOne* on a history-making flight just past the 100-kilometer (62.1 miles or 328,000 feet) line demarking the start of space. Now Rutan's team set their sights on the most exciting and influential prize of the new millennium.

Pete Siebold, who had already flown two glide flights and one powered fight in *SpaceShipOne*, was selected to pilot the flight. Siebold had been training for three years for this moment, but a health scare forced a very disappointing change.

"There were two other guys that were more than qualified to fly that flight," Siebold said. "At the time, I didn't feel as though I was doing the team any justice by putting myself in that situation and flying the mission when I probably wasn't in the right frame of mind and not to mention healthy enough." Siebold made the tough decision, but very fortunately his health issues were eventually determined to be nowhere near as serious as first suspected.

Rutan then turned to the test pilot with the most experience flying *SpaceShipOne*. Mike Melvill would get his chance to become an astronaut a second time, but to do so, he'd have to get back into training again. Figure 9.3 shows Melvill at the controls in the cockpit of *SpaceShipOne* preparing for X1.

Like *Spirit of St. Louis*, *SpaceShipOne* was stripped of anything absolutely nonessential. The lighter the craft, the greater the margin *SpaceShipOne* had for clearing the 100 kilometers (62.1 miles or 328,000 feet) because the removal of each and every pound enabled the spacecraft to go an additional 150 feet (46 meters) higher. *SpaceShipOne* needed all the help it could get. Melvill's earlier spaceflight had cleared the 100 kilometers by only the slimmest of

margins, less than 500 feet (150 meters). And during this spaceflight, *SpaceShipOne* was not even carrying the full payload required by the rules of the Ansari X Prize.

Ironically, as the Scaled Composites team scrimped for a pound here and a pound there, removing a total of about 45 pounds (20 kilograms), they would have to add weight to simulate two passengers. "We had to carry 400 pounds [180 kilograms] in the back seat, which was a heck of a lot more load in that thing than we ever had before. And I had to be ballasted," Melvill said.

Since Melvill weighed only 160 pounds (73 kilograms), he had to be ballasted up to 200 pounds (90 kilograms). These were the rules. But keeping the gross weight as low as possible was still critical. Every pound that didn't have to be carried was a pound that the force from the rocket engine didn't have to lift.

Figure 9.4 shows Melvill gesturing "okay" from a removable port as final preparations were made. *SpaceShipOne* was carried

Fig. 9.3. Pete Siebold was selected to fly the first Ansari X Prize flight. However, health concerns prevented him from flying. Mike Melvill, with more stick time in *SpaceShipOne* than Siebold and Binnie combined, was asked to fill in. Melvill had faced some demanding flights and had thought he was finished flying *SpaceShipOne*. He had been glad to be done, but he stepped up to the challenge and climbed back into the pilot's seat. *Mojave Aerospace Ventures LLC, photograph by Scaled Composites*

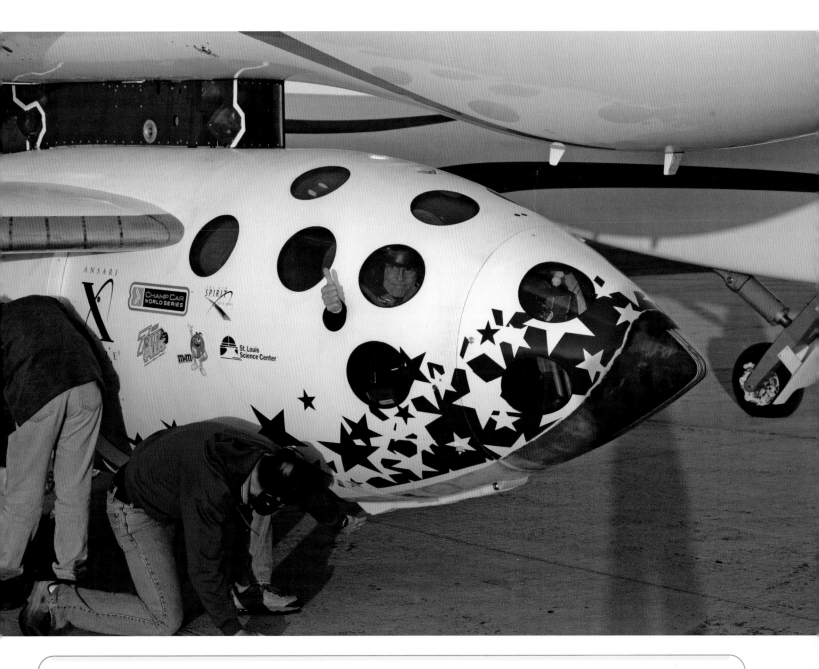

Fig. 9.4. Mike Melvill gives the thumbs up from the cockpit of *SpaceShipOne* as last-minute preparations are made. For the Ansari X Prize, *SpaceShipOne* had to carry enough weight to simulate three 198-pound (90-kilogram) people. *SpaceShipOne* just barely made it past 328,000 feet (100,000 meters) without the extra weight, so it was necessary to bump up the performance of the rocket engine. *Mojave Aerospace Ventures LLC, photograph by David M. Moore*

into the air at 7:12 a.m. PST by *White Knight* with Brian Binnie behind the controls.

Separation occurred at 8:10 a.m. PST when flight engineer Matt Stinemetze, who sat in the back seat of *White Knight*, released *SpaceShipOne* from an altitude of 46,500 feet (14,170 meters). Clear of *White Knight* and no longer pushing forward on the control stick, Melvill fired the rocket engine, which had been enhanced to provide greater performance by increasing the amount of propellant and burn time.

"You could sure hear it," Melvill said. "It was very loud—it was extremely loud.

"But it is a fabulous ride going up. I think that people who go on the next one—the passengers—will get the most exciting thing they ever did. A lot of noise. They are going really fast. The acceleration is dramatic. You are accelerating at a huge rate. You just watch the speed going up."

The cockpit shook as Melvill pulled the nose up, making a very sharp turn toward the sky above. "The straighter you flew it, the

Fig. 9.5. A video-capture image shot from a chase plane shows *SpaceShipOne* spiraling up during its boost. Melvill struggled to control the rolls but still allowed *SpaceShipOne*'s rocket engine to fire so he would be sure to pass the 328,000-foot (100,000-meter) mark. *SpaceShipOne* rolled twenty-nine times. *Mojave Aerospace Ventures LLC, video capture provided courtesy of Discovery Channel and Vulcan Productions, Inc.*

higher it would go in the same amount of time," he said. "We didn't need to burn the motor for its full length that it was capable of burning because it went up there quite easily."

During his previous flight, though, he had battled wind shear, rocket asymmetries, and pitch control failure. These had prevented him from flying a very straight trajectory. Melvill was more than determined to nail the trajectory on this flight.

As *SpaceShipOne* blasted through the upper atmosphere, Melvill had closed up the two donuts on the TONU display, which meant he was doing a great job flying the planned trajectory, and he monitored the energy altitude predictor, an instrument that predicted how high *SpaceShipOne* would go once the rocket engine was shut down.

"You may be at 160,000 feet [48,770 meters]," Melvill said, "and it will say, if you turn it off right now, you'll go all the way to 328,000 feet [100,000 kilometers]. So, you watch that instrument. That's the primary instrument to know when to turn it off. Initially, we did it with a timer, and we just said you're going to run 55 seconds. And at the end of 55 seconds, we'd shut it off."

But only 60 seconds after lighting the rocket engine, traveling at Mach 2.7, *SpaceShipOne* was in trouble. The crowd hushed as the contrail from *SpaceShipOne* switched from a nice, smooth line to a wild corkscrew in the sky. Things happened fast. But from the angle of the shot displayed on the jumbo screen, it was hard to tell what was actually happening. Figure 9.5 shows *SpaceShipOne* rolling out of control, viewed from the cockpit of a chase plane.

"When he started doing the rolls, I thought he was dead," recalled Erik Lindbergh. "I thought that was it—the craft was going to break up and he was done."

Thousands of people were gripped in silence.

"I didn't think he was doing rolls. I thought he was tumbling at that point," Lindbergh recalled.

SpaceShipOne rolled right uncontrollably at an initial rate of 190 degrees per second, spiraling up toward space.

"I had one of the walkie-talkies, and I could hear Melvill talking to ground control," Ansari said. "He said that everything is fine. It didn't look fine. But because he was convinced that everything was fine, I felt comfortable."

The rocket engine kept burning while *SpaceShipOne* still spun its way up, reaching a maximum speed of Mach 2.92 (2,110 miles per hour or 3,400 kilometers per hour). Melvill still kept his eye on the energy altitude predictor. As he explained, "Unless you see 328,000 feet [100,000 meters] in that window, you are not going to win the X Prize. So, you don't want to turn it off until you read at least that much or more. And so that was why I didn't turn it off when we were doing all those rolls, because it didn't say 328,000 feet [100,000 meters] yet. I went to turn it off thinking, wow, something was wrong here. And when I looked at the energy height predictor, it was not predicting that we would go high enough. So, I just left the motor running and just ignored the rolling."

At a total burn time of 77 seconds, Melvill finally shut off the rocket engine. His altitude was 180,000 feet (54,860 meters) at that point and only about halfway through its ascent. But as Melvill got higher and higher, the air became too thin for him to counteract the rolls with either the subsonic or supersonic flight controls. *SpaceShipOne* left the atmosphere still rolling at 140 degrees per second.

Melvill was able to keep from being disoriented by focusing on the Tier One navigation unit and not glancing out the windows. He activated the feather and then focused on nulling-out the rolls with the reaction control system. "I just pushed it on, turned on both systems, and just left it on until it stopped it," Melvill said.

By the time *SpaceShipOne* stopped rolling, it had completed twenty-nine rolls. The vehicle now continued to coast to an apogee of 337,700 feet (102,900 meters), but now Melvill could enjoy the 3.5 minutes of weightlessness and the view while still having time to take a few photos out the window.

On reentry, *SpaceShipOne* hit a top speed of Mach 3.0. Still in the feathered configuration, it decelerated from supersonic to subsonic,

Fig. 9.6. By using the reaction control system (RCS), Mike Melvill stopped the rolls. He reached an apogee of 337,700 feet (102,900 meters), which gave him about 10,000 feet (3,000 meters) to spare. Now, coming down was the easy part. *Mojave Aerospace Ventures LLC, video capture provided courtesy of Discovery Channel and Vulcan Productions, Inc.*

Fig. 9.7. *SpaceShipOne*'s second spaceflight was nearly over as it approached Mojave's runway. But before Melvill had gotten close to the airport, he did some early celebrating by rolling *SpaceShipOne* once more to make it an even thirty. *Mojave Aerospace Ventures LLC, video capture provided courtesy of Discovery Channel and Vulcan Productions, Inc.*

Fig. 9.8. Mike Melvill completed X1, the first Ansari X Prize flight. He did so with some unintended excitement, and that drew a little concern. *SpaceShipOne* had yet to fly a smooth and trouble-free spaceflight. Now Scaled Composites had two weeks to get *SpaceShipOne* ready for the second Ansari X Prize flight. *Mojave Aerospace Ventures LLC, photograph by David M. Moore*

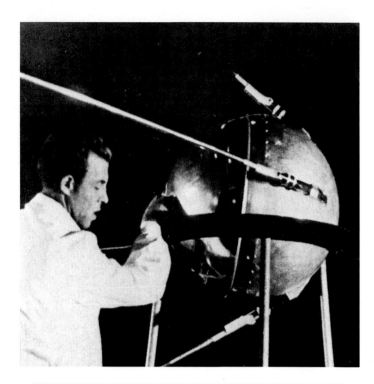

Fig. 9.9. On October 4, 1957, *Sputnik* became the first man-made object to orbit Earth. The Soviet Union had launched the beach ball-sized satellite, which circled the planet for three months. *Sputnik* put the space race between the United States and Soviet Union into overdrive. Decades later, on this day, *SpaceShipOne* was ready to finish a race of its own. *NASA*

while it reached a peak g-force of 5.1 g's at 105,000 feet (32,000 meters).

At 61,000 feet (18,590 meters), Melvill retracted the feather to begin his glide back to Mojave. As *SpaceShipOne* descended, the chase planes caught up and tucked in behind. Figure 9.6 shows the view of *SpaceShipOne* from the Alpha Jet.

During the 18-minute glide to Mojave, *SpaceShipOne* suddenly rolled completely around, surprising the chase planes. But this roll wasn't uncommanded. Melvill performed a victory roll, rounding out his total rolls for the flight at thirty.

Figure 9.7 shows *SpaceShipOne* coming in for a landing as the crowds lining the runway cheered.

"It was fabulous—it really was—knowing that we at least were halfway there. We went plenty high. And coming back and all the excitement, everybody was just thrilled to death," Melvill said.

Melvill's flight exceeded the altitude requirement by nearly 10,000 feet (3,050 meters) and satisfied the other rules set by the Ansari X Prize. To fulfill the remaining conditions, *SpaceShipOne* had to repeat the spaceflight within two weeks. Standing on *SpaceShipOne*, Melvill celebrated successfully completing the X1, as shown in figure 9.8.

"We knew what we had to do. My task was to not damage the airplane. I wasn't going to go for any altitude records but just plenty of margin and burn the engine as little as possible and land the airplane as smooth as possible so we didn't have to fix anything. We didn't even change the tires. We refueled it, and it was ready to go. We could have gone the next day," Melvill said.

Turnaround Time

The twenty-nine rolls not only caused great concern for safety, but now doubt and skepticism started to creep into the back of people's minds. One or two rolls wouldn't have been so dramatic, and would not have left such a vivid impression. But with twenty-nine, even the most inexperienced spectator could tell things weren't going as planned. The public had not yet bought into the whole idea of personal space travel. There was a big difference between being enthusiastic and thinking something was cool and being willing to put your own butt in the seat strapped to a rocket engine. Some people would of course be willing to take any risk to get into space. But that certainly wouldn't be the best way to jumpstart an industry in this day and age. Sometimes perception, unfortunately, weighs heavier than fact.

"We saw this rolling departure, and that was cause for concern," Binnie said. "Not from a safety or structural standpoint but a concern of perception. Others thought, 'Well, they are just loose cannons out there. They don't understand what they are doing. They are certainly not ready for prime time or carrying the trusting public.' And so the clock is ticking.

"We had planned this to where we could potentially pull off three flights in two weeks if need be. But we were all getting kind of tired. We really didn't want to have a problem on our second attempt. Everybody on the team was well aware of what was at stake and what would all be necessary should it have to come to a third flight. And any number of things could put us there. It could be bad weather, an avionics hiccup, range issues, telemetry things, and issues totally unrelated to flying the vehicle could have scuttled that event and forced us into a third evolution."

The fact was that after evaluating the data from X1, the team determined Melvill had done too good of a job at pointing *SpaceShipOne* straight up. In this orientation, *SpaceShipOne* had no aerodynamic lift to correct unwanted motion. "You've got to be careful that you don't go over on your back," Melvill said. "It is real easy to pull so hard that you end up overturning."

With the nose of *SpaceShipOne* pointed straight up, a degree or two off one way or the other was not much of a change in angle. But it turned out to be a tremendous change in terms of *SpaceShipOne*'s stability. So, when Melvill went beyond 90 degrees, he naturally tried to bring the nose back on track. This caused the angle of attack, the direction *SpaceShipOne* was pointing in relation to the direction of actual motion, to go to zero.

"The wing wasn't lifting anymore, there was zero lift on the wing, then it departed," Melvill explained. "It did a snap roll. And that was caused by the design of the airplane. The airplane was designed with a high wing and swept leading edge. Had that been a low wing, it would not have done what it did. We learned that lesson. On the next flight, we didn't change the airplane at all. We just changed the pull-up schedule."

The new plan for the trajectory was a more gradual pull-up during boost while making sure never to go to vertical. "And as long as that wing is lifting, it won't stall like that. But when it gets to zero lift, then you get separation on it, and the slightest little perturbation of airplane will cause it to roll or do something odd," Melvill said.

None of the flights previous to X1 had flown at high Mach numbers while at a zero angle of attack. Essentially, *SpaceShipOne* lost directional stability, so there was no way Melvill could counteract the weak thrust asymmetry, a wandering thrust line, coming from the rocket engine at the time. *SpaceShipOne* was still rocketing up, so by the time the first few rolls occurred, the atmosphere had disappeared. Aerodynamic forces were not longer causing the rolls, but since there was no air pressure to resist the rolling motion, once *SpaceShipOne* started to roll, it just kept going and going.

The structural loading on *SpaceShipOne* from the rolling was very low. Melvill's safety was never in jeopardy, only his breakfast, which thanks to all of the unusual-attitude training in the Extra 300 aerobatic plane, remained in place. The very next day, Scaled Composites not only figured out what caused the rolling departure but also determined a way to keep it from happening again.

As with the first rocket-powered launch of *SpaceShipOne*, Rutan wanted to fly on a significant day in aviation history. The anniversary of the first man-made object to orbit Earth was approaching. Russia's *Sputnik*, as shown in figure 9.9, was launched on October 4, 1957, and circled Earth about 1,400 times at a peak apogee of 588 miles (947 kilometers). This milestone of spaceflight sent the space programs of the United States and the USSR into warp speed.

"We had three days to finesse this in the simulator," Binnie said. "Between Mike's flight and the final flight, it was Friday, Saturday, Sunday. It looked promising, but it was still only our sixth powered flight in the vehicle. There was no guarantee that we really understood it or that there weren't some other gremlins that were going to leap out and get us."

X2: Winning the Ansari X Prize (17P)

On this day, the 47th anniversary of *Sputnik*, Brian Binnie was selected to pilot *SpaceShipOne* for the second of the two flights required to win the Ansari X Prize. Melvill, who had been a backbone of the program, had paved the way for Binnie only five days earlier. Melvill would now be there flying *White Knight* along with Matt Stinemetze as flight engineer.

B²

367,550' ── M = 3.26

x₁

Gone.

17P - Binnie

66L- Melvill/Stinemetze

Flight Test Log Excerpt for 17P

Date: 4 October 2004

	Flight Number	Pilot/Flight Engineer
SpaceShipOne	17P	Brian Binnie
White Knight	66L	Mike Melvill/Matt Stinemetze

Objective: Second X Prize flight: again ballasted for 3 place and 100 kilometer goal (328,000 ft). (We also really wanted to break the X-15 354 kft [thousand feet] record.)

(source: Mojave Aerospace Ventures LLC, provided courtesy of Scaled Composites)

"He dropped me, and I dropped him," Melvill said. "That was fun."

White Knight and *SpaceShipOne* lifted off together at 6:49 a.m. PST on October 4, 2004, in the chill of the desert morning with the Sun rising. In an article Binnie wrote for *Air & Space*, he echoed the thoughts of Melvill, "The program to develop and test Burt Rutan's *SpaceShipOne* (*SS1*) had many different demands, but I can safely say the one that made the pilots uniformly uncomfortable was the hour-long wait in *SS1* while the *White Knight* carrier aircraft dragged it up to release altitude. During this time, there is little to do and the mind is somewhat free to wander."

As the world watched, the pressure on Binnie was enormous. With the prize of $10 million on the line, Branson waiting down below poised to begin work on SpaceShipTwo, and the fact that it was ten months since the last time he flew *SpaceShipOne*, which resulted in a crash landing, Binnie had plenty to wrestle with inside his head. "For me personally, a problem or failure or inability to pull this off for whatever reason, the other side of that coin was a bottomless pit. It felt to me like an abyss."

Tensions ran high on the ground, too. "I knew that if there would be any glitches, people would say that this is not ready for prime time," Ansari said. "And it's not ready for commercialization and all these things."

But when it came time to launch, every trace of doubt or uncertainty disappeared with the first flash of the rocket engine. Binnie's years of navy flying and skills as test pilot took control. At 7:49 a.m., an hour after takeoff, and 47,100 feet (14,360 meters), Stinemetze pulled the lever to drop *SpaceShipOne*.

Table 9.1 gives the transcript of the communication between Binnie and Mission Control from the moments before separation to when the feather was locked down after reentry.

"We had no reason to delay," Binnie said. "So, as soon as I was separated, I armed and fired the rocket motor."

Ignition occurred immediately, and off Binnie went. *SpaceShipOne* zoomed past *White Knight* close enough for Melvill and Stinemetze to hear the hybrid rocket engine, a spaceman's version of buzzing the tower. Figure 9.10 shows *SpaceShipOne* beginning to make its turn toward space. After 10–12 seconds, Binnie was thinking, "Okay, I'm still alive. I'm still in the loop. I'm still managing this thing." But as *SpaceShipOne* transitioned into supersonic flight, he relaxed. The hardest part was over.

"We wanted to get to the X Prize altitude and a secondary goal of trying to beat the X-15 record," Binnie said. "So, we wanted lots of altitude. But we also wanted to exit the atmosphere without any rolls or gyrations or large body rates so that we didn't scare off Branson and the whole SpaceShipTwo efforts. There was that dual-edge sword of precision flying on one side and performance on the other.

"We wanted to get the nose up to 60 to 70 degrees as quickly as we could, initially, a very aggressive turn," Binnie continued. "Once we got there, we started slowing down the pitch rate on the vehicle so that we went from 60 degrees to about 80 to 82 degrees over the next 50 seconds or so. The bulk of the flight was just milking the nose between those pitch attitudes. And then the last 20 to 25 seconds was the start of a pull again to get to about 87 to 88 degrees nose up."

Fig. 9.10. Brian Binnie faced the toughest flight of his entire life on October 4, 2004. He hadn't flown *SpaceShipOne* since the crash landing. Now he was behind the controls of *SpaceShipOne* while the world's eyes watched his every move. *Mojave Aerospace Ventures LLC, video capture provided courtesy of Discovery Channel and Vulcan Productions, Inc.*

The exhaust from *SpaceShipOne*'s rocket engine streaking upward as the contrail from *White Knight* veers off to the left can be seen in figure 9.11.

Binnie continued, "The initial pitch attitude to 60 to 65 degrees meant you were going to take advantage of all that rocket motor energy that is available to you and convert that to altitude. And the pull-in endgame meant you were keeping angle of attack on the vehicle and making it less susceptible to rocket-motor asymmetry in the thin upper atmosphere, where you have a delicate balance between controlling those asymmetries with little aerodynamic control power to resist it."

After 84 seconds, Binnie shut down the rocket engine when *SpaceShipOne* had reached 213,000 feet (64,920 meters), zipping upward as fast as Mach 3.09 (2,186 miles per hour or 3,518 kilometers per hour). Like a pot of gold at the end of a rainbow, $10 million waited at the other end of the ballistic arc.

"I went scooting right through the X Prize altitude and past the X-15 old record by 13,000 feet [3,960 meters] or so. I got to the point after rocket motor shutdown and the feather coming up, and I hadn't touched any of the reaction control system yet to control body rates. The vehicle was just absolutely stable. I actually used reaction control to give myself a different view so I could take some pictures."

Fig. 9.11. The contrail of *SpaceShipOne* streaks spaceward as the contrail of *White Knight* peels off to the left. Brian Binnie fired the rocket engine for 84 seconds, shutting it down at 213,000 feet (64,920 meters). *Dan Linehan*

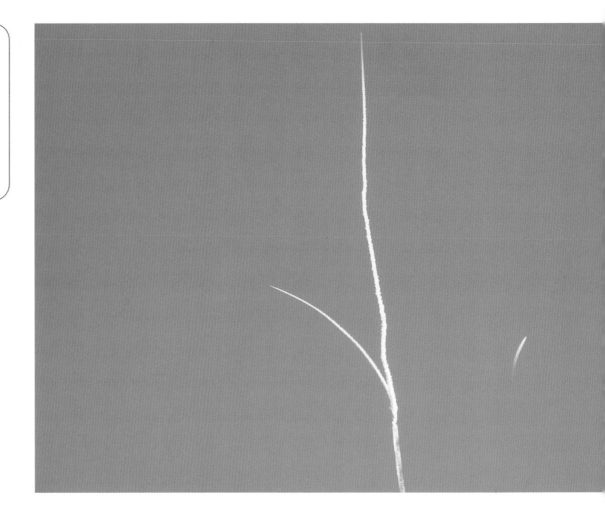

Fig. 9.12. Well past the boundary of space, Brian Binnie had entered the black sky. *SpaceShipOne* coasted to an apogee of 367,500 feet (112,000 meters), surpassing the X-15 altitude record of 354,200 feet (108,000 meters). *Mojave Aerospace Ventures LLC, photograph by Scaled Composites*

Binnie reached an apogee of 367,500 feet (112,000 meters), which exceeded the Ansari X Prize requirements by nearly 7.5 miles (12 kilometers), and experienced weightlessness for more than 3.5 minutes. Binnie had a little time to take pictures, and figure 9.12 shows one of his photographs. In addition to taking photos, as figure 9.13 shows, he had the chance to do some zero-g testing on a miniature *SpaceShipOne*. Binnie did not release M&Ms in space as did Melvill, and it's still unconfirmed whether Binnie ate his allotment during the captive-carry phase. Doug Shane would not speculate on the origin of several faint crackling sounds heard over the Mission Control radio.

Although weightless at apogee, *SpaceShipOne* had not truly escaped Earth's pull. *SpaceShipOne* started to freefall and began to

accelerate, reaching Mach 3.25, which was the fastest speed it had ever reached on any of the flights. As *SpaceShipOne* descended into the thick atmosphere, air friction now decelerated it, and at 105,000 feet (32,000 meters), Binnie faced a peak force of 5.4 g's pushing against his body.

As the g-forces subsided and *SpaceShipOne* slowed down below the speed of sound, Binnie retracted the feather at an altitude of 51,000 feet (15,540 meters). The video frames, at two-second intervals, in figure 9.14 show the transition of the feather mechanism from the extended position to the retracted position. After reentry into Earth's atmosphere, the feather had done its job. The pair of pneumatic actuators, which can be seen connecting either side of the fuselage to the trailing edge of the

Fig. 9.13. Brian Binnie's trajectory was spot-on during the ascent. The hard part was over, now that gravity had taken over control of the trajectory. Binnie had 3.5 minutes of weightlessness to savor. He snapped photos and sailed a *SpaceShipOne* model back and forth across the cockpit. *Mojave Aerospace Ventures LLC, video capture provided courtesy of Discovery Channel and Vulcan Productions, Inc.*

wings, pulled downward. This caused the feather to retract, making *SpaceShipOne* streamlined once more and readying it for the glide back to Mojave.

SpaceShipOne, now configured as a glider, drifted homeward. Figure 9.15 shows Sir Richard Branson, Paul Allen, and Burt Rutan spotting *SpaceShipOne* in the sky above Mojave.

The world watched *SpaceShipOne* gliding down for 18 minutes.

"I don't know," Ansari said, "maybe naively, I just felt that there was no more danger and everything would be fine or if there were any glitches or problems, they would be very much manageable. I wasn't too worried because I had watched landings of *SpaceShipOne* a few times before."

After only 24 minutes from being dropped by *White Knight*, *SpaceShipOne*'s wheels hit the runway for a perfect landing, as shown in figure 9.16.

"Oh, it was absolutely wonderful," said the 434th human to reach space, summing up his spaceflight.

Once the nose skid brought *SpaceShipOne* to a stop and the door popped open, Binnie was instantly welcomed back by his wife as Rutan, Allen, and Branson congratulated him on the victorious flight. Towed by a pickup truck, *SpaceShipOne* paraded up and down the flightline in front of the thousands and thousands of cheering supporters as Binnie stood triumphantly atop, as shown in figure 9.17.

"The whole experience was very emotional for me," Ansari said. "Even though I had nothing to do with the design and the hard work that the engineers and the team had put into building *SpaceShipOne*, I just felt like part of the team. I was just so proud and happy that they were successful, and that was the greatest joy to see that happen."

Eight years after it was announced, the Ansari X Prize was finally captured, just like the Orteig Prize, first offered in 1919 and claimed in 1927. The difference was that aviation would not just take a giant leap into the air but would leap past where the air was thin to the beginning of space.

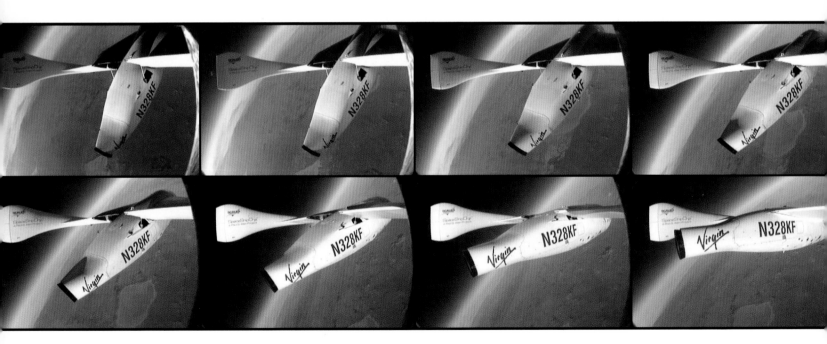

Fig. 9.14. As *SpaceShipOne* fell back to Earth, the feather eased it into the atmosphere. At 51,000 feet (15,540 meters), Binnie retracted the feather, as shown by the sequence given at two-second intervals. *Mojave Aerospace Ventures LLC, video captures provided courtesy of Discovery Channel and Vulcan Productions, Inc.*

Fig. 9.15. Sir Richard Branson, Burt Rutan, and Paul Allen (left to right), search the sky and spot *SpaceShipOne* as it nears the end of its 24-minute journey up and down from space. *X PRIZE Foundation*

Fig. 9.16. Brian Binnie finished his flawless performance by making a perfect landing. After about a year and a half of flight testing, seventeen flights altogether, *SpaceShipOne* touched down on the runway for the last time. *Mojave Aerospace Ventures LLC, video capture provided courtesy of Discovery Channel and Vulcan Productions, Inc.*

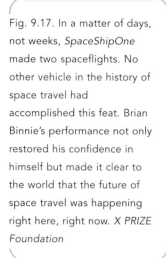

Fig. 9.17. In a matter of days, not weeks, *SpaceShipOne* made two spaceflights. No other vehicle in the history of space travel had accomplished this feat. Brian Binnie's performance not only restored his confidence in himself but made it clear to the world that the future of space travel was happening right here, right now. *X PRIZE Foundation*

Figure 9.18 shows Peter Diamandis and Anousheh Ansari celebrating with Brian Binnie, Burt Rutan, Paul Allen, and Sir Richard Branson. "It was just a feeling of relief that everything had worked flawlessly," Paul Allen recalled. "A mix of elation and relief I think is what I described at the time. And you are proud for Burt and his team. In the back of your mind you are thinking like maybe this does open the door for a lot of possibilities in the future in terms of private space tourism. I was just very excited and relieved, just an amazing mixture of emotions."

With the Ansari X Prize awarded, commercial space travel officially launched off. Diamandis' vision of a new way of thinking about space flight became reality, and Rutan with his team from Scaled Composites provided the way. Eight years was a long race, but the accomplishments during this time frame far outreached what was once thought possible.

The *will* was now strong enough to overcome the energy barrier to space, much the way the mystical sound barrier was broken in the 1940s to usher in supersonic flight. Paul Allen saw to it that Burt Rutan would have the chance to show his stuff and prove to the world that the impossible wasn't impossible. And Brian Binnie's perfect performance flying *SpaceShipOne*, gave all the reason to Sir Richard Branson and his newly formed Virgin Galactic that commercial space travel was right.

"Burt has the world's greatest garage," Paul Allen said. "We built a rocket in the world's greatest garage, and we actually got into space and back, and everybody was safe. And it won a prize. It is hard to explain the excitement of that. And the crowds being there celebrating that with you was just amazing."

Fig. 9.18. *SpaceShipOne* had done it. Eight years after its announcement by Peter Diamandis and the X Prize Foundation, Brian Binnie had captured the Ansari X Prize. Burt Rutan, Paul Allen, and the rest of their team had pulled off the seemingly impossible. Now was the time to celebrate the historic accomplishment and also to revel in the wonderment as the door to space flung wide open. *X PRIZE Foundation*

Table 9.1 Transcript of *SpaceShipOne's* Ansari X Prize–Winning Spaceflight

This transcript was prepared using video filmed during the second Ansari X Prize spaceflight attempt from inside the cockpits of *SpaceShipOne* and *White Knight* and from inside Mission Control. The spaceflight was called X2 because it was the second attempt required by the Ansari X Prize and also called 17P because this rocket-powered flight was the seventeenth time *SpaceShipOne* flew. *White Knight* lifted off with *SpaceShipOne* from Mojave's Runway 30 at 6:49 a.m. PST on October 4, 2004. The entire spaceflight lasted 1.6 hours (wheels up to wheels down for *White Knight*). The time stamps are hours:minutes:seconds a.m., PST. The transcript runs from just before *SpaceShipOne* is dropped from *White Knight* to just past feather retraction and lock after reentry. Acronyms used are:

AFFTC:	Air Force Flight Test Center at Edwards Air Force Base
AST:	Federal Aviation Administration Office of Commercial Space Transportation
BB:	Brian Binnie in *SpaceShipOne*
BR:	Burt Rutan in Mission Control
DS:	Doug Shane in Mission Control
MC:	Staff in Mission Control
MM:	Mike Melvill in *White Knight*
MS:	Marc de van der Schueren in the Alpha Jet chase plane
SS1:	*SpaceShipOne*

Time stamp and speaker	Dialogue	Notes
7:49:17 MM	"SCUM status?"	The Scaled Composites Unit Mobile (SCUM) was a mobile ground control station.
7:49:19 DS	"SCUM is go for release and ignition, elevons to go."	BB pushes the control stick forward, preparing for release.
7:49:22 MM	"Three. Two. One."	
7:49:25 MM	"Release."	
7:49:27 BB	"Release."	
7:49:28 BB	"Armed."	
7:49:28 BB	"Fire."	
7:49:32 BB	"Good light."	
7:49:37 DS	"Coming up ten seconds, Brian."	Rocket engine burn time at 10 seconds.
7:49:39 BB	"Copy."	
7:49:42 DS	"Rates look good and low."	
7:49:46 DS	"Okay, start the nose-down trim."	The trajectory must be timed in order to end up at zero angle of attack as late as possible.
7:49:49 DS	"Looking great at twenty seconds."	Rocket engine burn time at 20 seconds.
7:49:54 DS	"Doing okay?"	
7:49:55 BB	"Doing alright."	

(Source: Mojave Aerospace Ventures LLC, provided courtesy of Scaled Composites)

Time	Transcript	Note
7:49:56 BB	"A little lateral oscillations now."	
7:49:58 DS	"Copy that. Thirty seconds, a little nose up trim is probably okay now."	Rocket engine burn time at 30 seconds.
7:50:02 MM	"You look great."	
7:50:07 BB	"Starting to settle out."	
7:50:09 DS	"Okay, forty seconds."	Rocket engine burn time at 40 seconds.
7:50:10 DS	"The energy's on the line. The trajectory looks good."	The actual trajectory is tracking with the predicted trajectory.
7:50:13 BB	"Copy."	
7:50:16 DS	"Touch of nose up trim."	
7:50:19 DS	"Fifty seconds."	Rocket engine burn time at 50 seconds.
7:50:21 DS	"Two hundred energy."	This reading stands for a projected altitude of 200,000 feet (60,960 meters) for *SS1*. It does not stand for *SS1*'s actual altitude. Both MC and AFFTC use energy altitude predictors to project the maximum altitude *SS1* would reach if its rocket engine were to shut off and *SS1* were to coast the remainder of the way up. The projected altitude gives an indication at any given time whether or not *SS1* will reach the target altitude of 328,000 feet (100,000 meters).
7:50:22 DS	"A little right roll trim."	A slight correction is made to the trajectory.
7:50:26 DS	"Nose pitch up, Brian, nose up trim."	
7:50:30 BB	"There is the shaking."	The liquid to gas transition, which occurs as the N_2O begins to run low in the oxidizer tank, causes this to happen.
7:50:31 DS	"Okay."	
7:50:33 DS	"Roll right."	
7:50:36 DS	"Three hundred thousand."	The predicted altitude is 300,000 feet (91,440 meters).
7:50:41 DS	"Three twenty-eight."	The predicted altitude is 328,000 feet (100,000 meters).
7:50:44 DS	"Radar is three twenty-eight."	The predicted altitude is 328,000 feet (100,000 meters) as measured by AFFTC.
7:50:45 BB	"Copy that."	
7:50:48 DS	"Three fifty suggest shutdown."	The rocket engine is still firing, and if it is shut down at this point, *SS1* will coast to 350,000 feet (106,700 meters).
7:50:53 BB	"Roger. Shutdown."	BB lets the rocket engine burn an extra 4–5 seconds.
7:50:58 BB	"And the rates look good."	
7:51:00 DS	"Okay. Copy that."	
7:51:01 DS	"You are going to want to track north for the entry."	A box, approximately 2.5 square miles in size, is set by AST for *SS1* to make the reentry.
7:51:03 DS	"You are just across the orange line on the south."	The orange line is an AST boundary.

7:51:05 DS	"Uh, you are good east, uh, east-west of Mojave."	
7:51:08 BB	"Okay, I see that."	
7:51:10 DS	"Okay."	
7:51:14 DS	"Eighty-four seconds."	The rocket engine burns for a total time of 84 seconds.
7:51:15 DS	"Eighty-four seconds, the shutdown is clean and the feather is green."	
7:51:19 BB	"Feather unlock."	
7:51:24 BB	"Feather . . . moving."	
7:51:27 BB	"RCS on."	The reaction control system (RCS) controls motion of *SS1* in space.
7:51:28 DS	"Copy that, Brian. It's moving, and it's green."	The feather is extending upward.
7:51:30 DS	"CTN is a little warm but looking fine."	The CTN is the case/throat/nozzle assembly of the rocket engine.
7:51:33 DS	"RCS A looks nominal."	The pressure for the RCS looks good.
7:51:37 BB	"I show the feather up."	
7:51:39 DS	"We do show the feather all of the way up now. It is green."	
7:51:48 BB	"The trim is set."	
7:51:49 DS	"Looks great."	
7:51:51 BB	"And I'm upside down."	
7:51:53 DS	"Okay."	
7:51:57 DS	"You are going to want to orient northwest for the entry, Brian."	*SS1* should be oriented so that it points toward Mojave.
7:52:00 BB	"Okay, Doug. Copy that, northwest"	
7:52:02 DS	"Sound great. Feel good?"	
7:52:04 BB	"I'm feeling great."	
7:52:05 DS	"Copy that."	
7:52:07 BB	"Better get the camera out."	
7:52:09 DS	"Roger that."	
7:52:14 BR	"X-15 record."	This comment is made in MC and not heard over the radio. *SS1*'s actual altitude is above the highest altitude reached by the X-15. The predicted altitude is no longer used.
7:52:16 DS	"X-15."	
7:52:21 BB	"Outstanding."	
7:52:22 MM	"That's outstanding. I knew it."	

7:52:37 BR	"Ten thousand feet over X-15."	This comment is made in MC and not heard over the radio.
7:52:48 BB	"Boy, it's really quiet up here."	
7:52:59 DS	"Okay, flight is through this position coming downhill through three fifty."	The actual altitude of *SS1* is 350,000 feet (106,700 meters).
7:53:03 DS	"And current position is five southwest."	
7:53:06 DS	"Correction, five south of the bull's-eye."	The bull's-eye is the center of the AST box.
7:53:08 DS	"Looks like the entry point is between main base and north base."	This is to let the chase planes know that reentry will occur between Mojave Air and Space Port and Edwards Air Force Base.
7:53:17 DS	"Brian, if you can keep it upright for GPS, that's good."	
7:53:19 DS	"And, again, orient northwest please for the entry."	
7:53:25 BB	"Copy that, Doug."	
7:53:36 DS	"And Brian, a little blip of right yaw trim would be good."	
7:53:43 DS	"That looks great."	
7:53:45 DS	"Doing okay?"	
7:53:48 BB	"I'm doing great, Doug. Uh, camera is, uh, stowed again."	
7:53:54 DS	"Copy that, passing two six zero."	The actual altitude is 260,000 feet (79,250 meters).
7:54:03 BB	"And it's northwest you want for the heading right?"	
7:54:06 DS	"Affirmative."	
7:54:09 DS	"That'll point you back at high key."	High key is a glide mode of the TONU.
7:54:21 DS	"All systems are green here, Brian."	
7:54:22 DS	"Don't worry about temps in the back end."	
7:54:24 DS	"We're looking good here."	
7:54:26 MS	"Alpha's got a visual."	The Alpha Jet chase plane spots *SS1*.
7:54:27 BB	"Okay, here comes the g's."	
7:54:29 DS	"Copy that."	
7:54:31 DS	"One hundred fifty thousand."	The actual altitude is 150,000 feet (45,720 meters).
7:54:37 BB	"There's three."	BB is referring to the number of reentry g's.
7:54:41 DS	"Max Mach is past three two six."	*SS1* reaches a maximum of Mach 3.25.
7:54:45 BB	"Five g's."	
7:54:58 DS	"Peak g is passed."	

7:55:00 BB	"Copy that."	
7:55:02 DS	"You're looking great on glide range."	
7:55:03 DS	"Coming through seventy five thousand."	The actual altitude is 75,000 feet (22,860 meters).
7:55:12 DS	"Ok, we have had no GPS."	
7:55:13 DS	"So, you are a little higher than your indication Brian."	
7:55:15 DS	"We are showing seventy but radar is looking like seventy-five now."	DS is reporting altitude in thousands of feet.
7:55:20 DS	"And, uh, roll right if you can, that would be good."	
7:55:23 BB	"Okay."	
7:55:27 DS	"Okay, there is seventy thousand radar."	The actual altitude is 70,000 feet (21,340 meters).
7:55:29 DS	"Feather at your discretion."	
7:55:31 DS	"Uh, might give it another couple seconds."	There is no hurry, as BB is in good glide range, and the vehicle should fly better.
7:55:32 BB	"It feels a little, ah, loosy goosy right now."	
7:55:35 DS	"Copy that."	
7:55:40 DS	"You are going to want to get the roll trim back to neutral as you defeather."	
7:55:45 DS	"Radar altitude sixty-three now, sixty-three thousand."	The actual altitude is 63,000 feet (19,200 meters).
7:55:48 BB	"Okay, I feel good about the feather."	
7:55:50 DS	"Yeah, we do here."	
7:55:55 DS	"RCS off when you can."	
7:55:58 BB	"RCS is off."	
7:55:59 DS	"Copy."	
7:56:00 DS	"You are going to want to start a right turn to the north as soon as you recover."	
7:56:04 BB	"Copy."	
7:56:08 DS	"Radar shows fifty-four thousand."	The actual altitude is 54,000 feet (16,460 meters).
7:56:17 BB	"I show the feather locked."	
7:56:18 MC	Cheering in mission control.	
7:56:19 DS	"The feather is locked and it is green."	

SpaceShipOne's journey had not ended with the Ansari X Prize, although the mission and the destination had substantially changed. *SpaceShipOne* was about to embark on two of its longest flights ever, one across the United States and one across the Solar System. *Tyson V. Rininger*

Chapter 10

Science Fiction to Science Fact

On November 6, 2004, the X Prize Foundation presented the Ansari X Prize trophy and the $10 million. Figure 10.1 shows Burt Rutan, Paul Allen, Mike Melvill, and Brian Binnie with members of the X Prize Foundation, Peter Diamandis, Gregg Maryniak, Amir Ansari, and Bob Weiss holding up the prize money. In order to also join in the celebration, Allen had flown the entire Scaled Composites team in one of his private airliners to the award ceremony held in St. Louis. Figure 10.2 shows the Scaled Composites team from an earlier photograph.

SpaceShipOne and the Ansari X Prize began on two separate but parallel courses. When they converged, their combined importance was greater than the sum of the two parts. It is difficult to imagine what the result would have been if *SpaceShipOne* or the Ansari X Prize had been taken out of the equation. Would another team have won the Ansari X Prize with the deadline and the funding set to expire in just a few months? Would the general public have had the awareness or been as involved to the degree that it was without the Ansari X Prize? Without the space mania would investors like Sir Richard Branson have embraced Rutan with such a sizable financial commitment?

The years 1996 to 2004 were very much a different time compared to the years 1919 to 1927. And although the Ansari X Prize was modeled after the Orteig Prize, it certainly was not a one-to-one substitution. At the end of the day, the X Prize Foundation did what they had to do to realize their dream. At the end of the day, Scaled Composites did what they had to do to realize theirs.

Fig. 10.1. The Ansari X Prize trophy and $10 million check were presented on November 6, 2004, to Mojave Aerospace Ventures, the official partnership between Paul Allen's Vulcan and Burt Rutan's Scaled Composites. The photograph shows Bob Weiss, Gregg Maryniak, Amir Ansari, Peter Diamandis, Brian Binnie, Mike Melvill, Burt Rutan, and Paul Allen (left to right) at the award ceremony hosted in St. Louis. *Mojave Aerospace Ventures LLC, photograph by Scaled Composite*

Fig. 10.2. In the Mojave Desert, which is referred to as the birthplace of the sonic boom, Scaled Composites, a small company founded by Burt Rutan in 1982, grew from an innovator in aircraft to an innovator in spacecraft. Without the efforts of the whole team, *SpaceShipOne* would never have been able to burst through Earth's atmosphere and truly become a spaceship. *Mojave Aerospace Ventures LLC, photograph by Scaled Composites*

Fig. 10.3. Influenced by the success of the Ansari X Prize, NASA announced the Centennial Challenges in 2005. John Carmack's Armadillo Aerospace, an Ansari X Prize competitor, just missed winning the Lunar Lander Challenge in 2006. The lander, shown here, demonstrated vertical takeoff, hover, horizontal translation, and vertical descent, but it couldn't stick the landing in the end. *Dan Linehan*

It is safe to say that those who have dreamed of someday flying into space had their chances become much, much better because of Rutan and the Ansari X Prize, whether it be on a ride in SpaceShipTwo with Virgin Galactic or in another suborbital spacecraft from a different spaceline.

Beyond the Ansari X Prize

As the involvement and development of the commercial space industry continues to move forward and expand, many new ideas and designs are being introduced to the public. Even NASA has gotten into the spirit of public and commercial spaceflight. In 2005, the agency announced the first two cash prizes in a series called Centennial Challenges: the space tether and beam-power challenges,

which are both the components needed to build an elevator to space. In 2006, the less obscure lunar lander challenge was added, and other challenges soon followed.

NASA has partnered with the X Prize Foundation to run some of the Centennial Challenges during the annual X Prize Cup. Figure 10.3 shows the lunar lander of John Carmack's Armadillo Aerospace, a team that had competed for the Ansari X Prize, whose amazing demonstration missed winning the challenge in 2006 by the slimmest of margins.

The X Prize Cups are a cross between air shows and space expos, where companies show off and, in some cases, even demonstrate many of the latest and greatest ideas. One of the big attractions is the Rocket Racing League, which is still in development.

Fig. 10.4. Introduced in 2005, the annual X Prize Cup showcases many of the new ideas and developments in the commercial space industry. The X Prize Cup also hosts several of NASA's Centennial Challenges. As the technology progresses, the X Prize Cup plans to host rocket-powered races. The photograph above shows a prototype rocketplane for the Rocket Racing League. *Dan Linehan*

Sean D. Tucker, a champion aerobatic pilot who is looking forward to flying in the league, said, "It is going to be very similar to a Red Bull course except longer and higher, and I think there are going to be milestones in the sky and altitudes you have to hit in the sky as well and then come back down. I think it is going to be a truly three-dimensional course, They're working with the technology now to have heads-up displays where you can see the virtual course in the air." Figure 10.4 shows a prototype rocket racer.

As recently as 2001, Dennis Tito became the first paying space tourist, flying to the *International Space Station* aboard a Russian Soyuz. Since then, four others have made this $20 million, or more, journey. In 2006, Anousheh Ansari was doing research based upon

this type of spaceflight for a venture she was involved with. She said, "I was looking into it to find out what type of training was really required if we were to commercialize orbital flights. Do people really need six months of training and all these things? The best way to find out was to go through the program. I started training as a backup. But three weeks before the flight the primary crewmember got ill. He failed one of his medical tests. And that's when they said, 'Well, if you want to go, you can go. You can take that seat now.' And I just couldn't say no to that."

Figure 10.5 shows Ansari floating about in the *International Space Station*. She skipped over suborbital entirely and went straight to orbital. With Ansari's support, she helped open the door to space a

Fig. 10.5. Space tourism began in 2001 when Dennis Tito rocketed to the *International Space Station* in a Russian Soyuz. In 2006, Anousheh Ansari joined the handful of people who have made this same journey. At between $20 million and $40 million, this ticket is out of reach from most people. But a growing number of entrepreneurs are recognizing that there is not just a desire for space but a demand for space. *Prodea Systems, Inc. All rights reserved. Used under permission of Prodea Systems, Inc.*

little wider for the rest of the public. This unexpected opportunity for her was well deserved.

One of the bigger prizes still out there is the $50 million America's Space Prize announced in 2004 by Bigelow Aerospace, which is an orbital version of the Ansari X Prize. In 2007, the X Prize Foundation raised the ante, not in terms of money but in terms of miles. Partnering with Google, the $30 million Google Lunar X Prize will have teams compete to land on the Moon. This is a one-way ride, though. No self-replicating, carbon-based life forms are required for the trek. But before orbital or lunar spaceflights get going for the public, there is still another race on for suborbital flights. About a dozen companies are currently developing suborbital spacecraft, several of which were Ansari X Prize competitors, like Starchaser and the da Vinci Project. The truth is, there is an enormous amount of activity behind the scenes as well as on center stage.

SpaceShipTwo and SpaceShipThree

Based on the design and trajectory of *SpaceShipOne*, now a proven space-ship, SpaceShipTwo takes advantage of the lessons learned while flying *SpaceShipOne*. Rutan stated his commitment to making it one hundred times safer than anything that has previously carried people to space.

On September 27, 2004, just days before the first Ansari X Prize flight attempt, entrepreneur Sir Richard Branson, founder of Virgin Records and Virgin Atlantic, entered into an agreement with Paul Allen and Burt Rutan to build a fleet of SpaceShipTwos to be launched from carrier aircraft similar to *White Knight*. SpaceShipTwo is about three times the size of *SpaceShipOne,* and its carrier is as large as an airliner. Figure 10.6 shows a conceptual drawing of SpaceShipTwo and its carrier aircraft, and figure 10.7 shows a size comparison that includes *SpaceShipOne* and SpaceShipTwo.

Branson formed the spaceline Virgin Galactic, in which he designated the first SpaceShipTwo the Virgin SpaceShip (VSS) *Enterprise* after *Star Trek*'s famed spaceship, and the carrier aircraft *Eva* after his mum. Virgin Galactic will pay $250 million for a fleet containing five SpaceShipTwos and two White Knight Twos. True to Rutan fashion, the program to develop these aircraft, called Tier 1b, is top secret.

The trajectory is a basic up and down, like *SpaceShipOne*'s, but initial launches will likely take place where the carrier aircraft flies from Mojave out over the Pacific Ocean, as shown in figure 10.8. SpaceShipTwo then reenters the atmosphere after having reached a reported apogee of 84–87 miles (135–140 kilometers), whereas during *SpaceShipOne*'s final flight it hit 69.6 miles (112 kilometers), well above the Ansari X Prize limit of 62.1 miles (100 kilometers). Having a better glide range than its predecessor, SpaceShipTwo will take a scenic glide back to Mojave.

SpaceShipTwo will have two pilots and room for six passengers. The price for the 2.5-hour trip will be about $200,000 to start with, which includes an orientation flight where the passengers on deck get to watch an actual flight of SpaceShipTwo from the carrier aircraft. That hefty price tag will come down once the economy of scale and competition begin to take hold. Virgin Galactic released mockups of the interior in September 2006. As shown in figure 10.9, passengers

Fig. 10.6. An early conceptual design of SpaceShipTwo and White Knight Two shows the similarities to their predecessors. However, SpaceShipTwo will be three times the size of *SpaceShipOne* and has a cabin the size of a Gulfstream 4 corporate jet while White Knight Two will be larger than a Boeing 757. Virgin Galactic's initial fleet will include five SpaceShipTwos and two White Knight Twos. *Courtesy of Virgin Galactic*

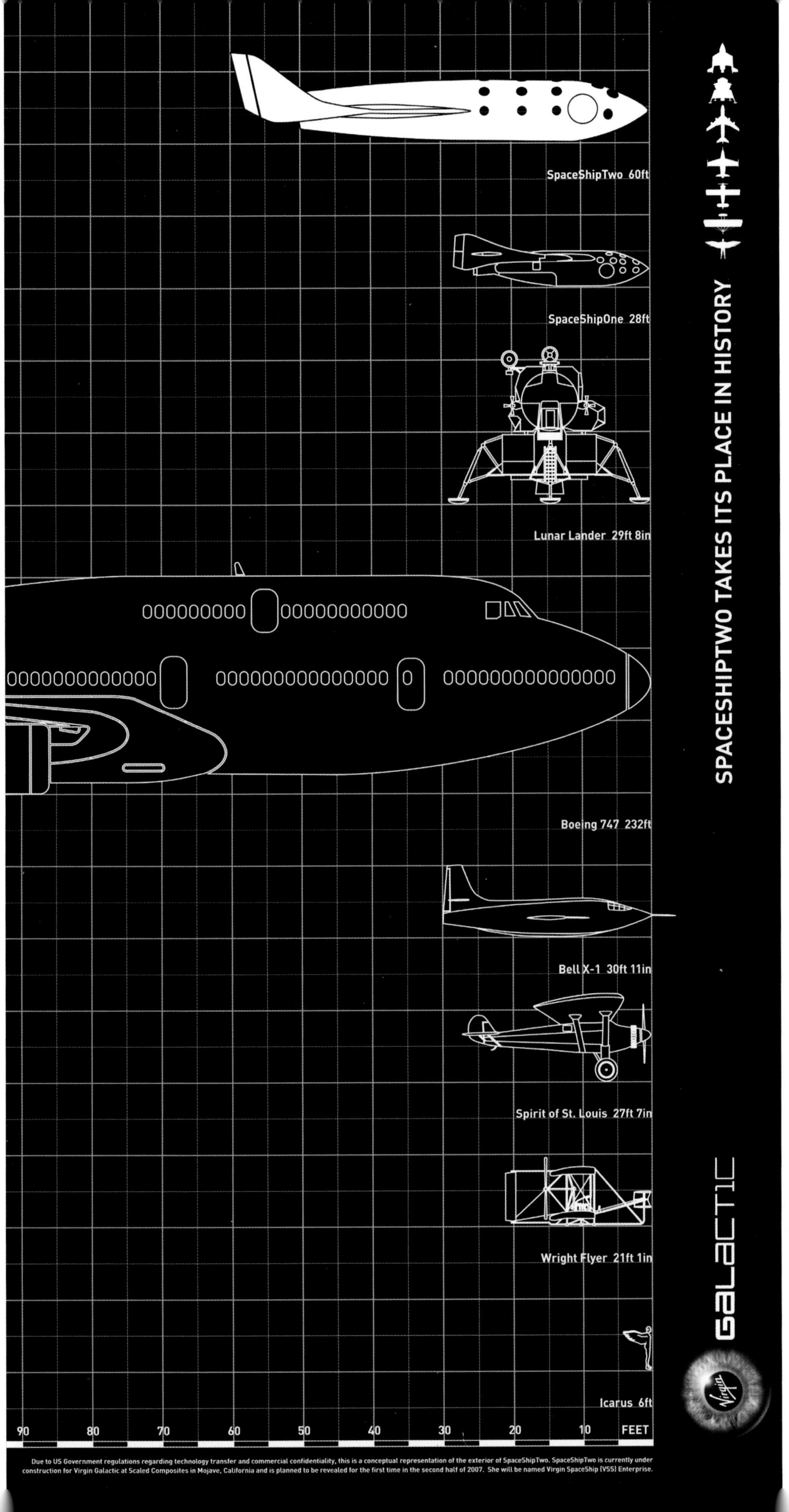
SpaceShipTwo 60ft
SpaceShipOne 28ft
Lunar Lander 29ft 8in
Boeing 747 232ft
Bell X-1 30ft 11in
Spirit of St. Louis 27ft 7in
Wright Flyer 21ft 1in
Icarus 6ft
90 80 70 60 50 40 30 20 10 FEET
SPACESHIPTWO TAKES ITS PLACE IN HISTORY
GALACTIC
Virgin
Due to US Government regulations regarding technology transfer and commercial confidentiality, this is a conceptual representation of the exterior of SpaceShipTwo. SpaceShipTwo is currently under construction for Virgin Galactic at Scaled Composites in Mojave, California and is planned to be revealed for the first time in the second half of 2007. She will be named Virgin SpaceShip (VSS) Enterprise.

Fig. 10.7. At an initial price of $200,000, a ride on SpaceShipTwo will cost a lot of spacebucks. But when bicycles and automobiles were first invented, not many people could afford them. And when ocean liners started to sail and airliners started to fly, the tickets were well beyond the reach of most. So, the ticket price of SpaceShipTwo is expected to drop substantially once competition between other spacelines takes hold and the space tourism industry begins to mature. *Courtesy of Virgin Galactic*

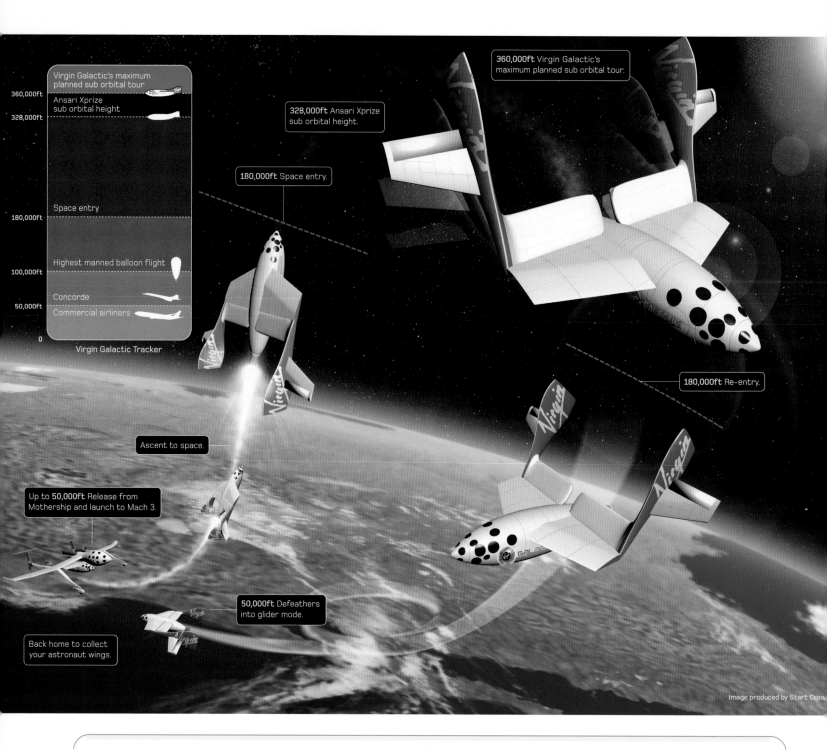

Labels within figure:

Virgin Galactic Tracker
- 360,000ft — Virgin Galactic's maximum planned sub orbital tour
- 328,000ft — Ansari Xprize sub orbital height
- 180,000ft — Space entry
- 100,000ft — Highest manned balloon flight
- 50,000ft — Concorde / Commercial airliners
- 0

360,000ft Virgin Galactic's maximum planned sub orbital tour.

328,000ft Ansari Xprize sub orbital height.

180,000ft Space entry.

180,000ft Re-entry.

Ascent to space.

Up to **50,000ft** Release from Mothership and launch to Mach 3.

50,000ft Defeathers into glider mode.

Back home to collect your astronaut wings.

Image produced by Start Crea

Fig. 10.8. This conceptual diagram shows an early representation of SpaceShipTwo's flight profile based on the flight profile of *SpaceShipOne*. SpaceShipTwo will be three times as large as its predecessor but will share many of the same design elements. *Courtesy of Virgin Galactic*

will be able to release their seatbelts and float around inside the cabin during a weightlessness period of about four minutes.

The conceptual drawings and early flight specifications available to the public will undoubtedly differ a bit from the end results. The launch altitude of 60,000 feet (18,290 meters) that has been floating around, for example, is a number that will likely come down. Just because the vehicles are larger and apogee is planned to be higher doesn't necessarily mean everything else scales up, too. After all, *SpaceShipOne* was planned to launch at 50,000 feet (15,240 meters), dropped to 48,000 feet (14,630 meters), and then finally ended up

Fig. 10.9. A mockup of SpaceShipTwo's cabin interior was revealed in 2006. Six passengers will ride to space, and when they get there, they will be able to unbuckle their seatbelts and float around the cabin to enjoy the weightlessness and the view. *Courtesy of Virgin Galactic*

at 47,000 feet (14,330 meters). To turn the corner with that big rocket engine blaring, SpaceShipTwo still needs air for the control surfaces to bite into, right? Will SpaceShipTwo even be mounted underneath the carrier aircraft, or will it ride on top like the Space Shuttle on a 747? Why risk your spaceship if your carrier aircraft has a landing gear failure? Why waste energy pulling downward away from the carrier aircraft during separation? However, a carrier aircraft could fly to a higher altitude in order to reduce the fuel requirements of a top-launching spacecraft if the carrier aircraft was able to pitch up and begin turning the corner for the spacecraft prior to separation.

SpaceShipTwo is set to fly passengers in the 2008–2009 timeframe. But before flight testing begins, SpaceShipTwo will be unveiled. SpaceShipThree will eventually follow. It will be the first of Rutan's Tier Two vehicles, designed for Earth orbit. SpaceShipThree certainly has a model number by now.

Two Last Flights for *SpaceShipOne*

Scaled Composites received about $25 million from Paul Allen for twenty tasks that Burt Rutan had specifically outlined, which covered building *SpaceShipOne* all the way through competing with it. "Task 21 was that we would fly *SpaceShipOne* every Tuesday for five months, reasoning that if we did that you could then make with confidence a commercial business plan," Rutan said.

But Task 21 wasn't funded. Rutan figured that once he got the data on the real costs of flying *SpaceShipOne*, he would then approach Allen. "That would be the opportunity for Paul and me and both of our friends to be astronauts," Rutan explained. "If you just count only the passengers, you've got forty-four people. So, maybe twenty of my friends could be astronauts and twenty of his friends could be astronauts. That would be kind of cool. That was the plan. But something got in the way of the plan. I underestimated the impact of *SpaceShipOne* on the media and the public, and I underestimated its effect on historians."

Shortly after Melvill flew *SpaceShipOne* into space the first time, Rutan received a letter from Valerie Neal, the curator of post-Apollo human spaceflight for the Smithsonian Institution's National Air and Space Museum. "It was clear to all of us right away once Mike Melvill had made the first flight in June that this was a remarkable achievement, whether or not it won the Ansari X Prize," Neal said.

"We think that *SpaceShipOne* either itself may prove to be the pivotal craft that leads to a commercial spaceflight space-tourism industry, or it's the leading edge of that. You know there are enough developments going on right now. It looks as if this is the cusp of a new revolution in spaceflight."

So, the National Air and Space Museum expressed its interest in acquiring *SpaceShipOne* to join it with other remarkable vehicles in the Milestones of Flight gallery, which includes the original 1903 Wright Flyer, *Spirit of St. Louis*, and the Bell X-1 that broke the sound barrier, *Glamorous Glennis*. But the National Air and

Fig. 10.11. When *SpaceShipOne* made it first spaceflight on June 21, 2004, the National Air and Space Museum of the Smithsonian Institution immediately recognized the significance of the event. By becoming the first non-governmental, privately funded vehicle to reach space, *SpaceShipOne* earned a place in the Milestones of Flight gallery with the *Spirit of St. Louis*, Bell X-1, 1903 Wright Flyer, and Apollo 11 command module *Columbia*. *Courtesy of Virgin Galactic*

Fig. 10.10. On June 27, 2005, Burt and Tonya Rutan, in *SpaceShipOne*, and Mike and Sally Melvill, in *White Knight*, landed at Oshkosh, Wisconsin, for the Experimental Aircraft Association's (EAA) 2005 AirVenture. An active EAA member, Burt Rutan introduced the VariViggen, the first aircraft he designed and built, at the 1972 AirVenture. Now he and Melvill, also a longtime EAA member, gave a special showing of *SpaceShipOne* and *White Knight* to many of their closest supporters. *Tyson V. Rininger*

Space Museum didn't realize the extent to which Allen was involved. Rutan said that he would have to bring Allen into the dialog as well.

"So," Neal recalled, "it was right at the end of November or early December when Allen and Rutan both said, 'Yeah, we're really interested in donating this to the museum. Come on out and let's talk and let's have a look at it together.'"

Rutan had to face a tough decision. He explained, "When we got that request, Paul Allen called and said, 'Listen, I don't want you to fly it anymore. Just get the X Prize. Two more flights and

Fig. 10.12. Carrying a small piece of *SpaceShipOne*, the space probe *New Horizons*, launched in 2006, races to the edge of the Solar System. The first mission ever to the dwarf planet Pluto, it will arrive in 2015. *NASA/Johns Hopkins University Applied Physics Laboratory–Southwest Research Institute*

that's it.' We had three or four motors, so we could have easily flown one more flight. My first thought was to fight with him. I said, 'No, you've got to prove a business plan. If this is going to go on to the next step, you got to do this.' And then I realized that he really was right."

Preserving the legacy was more important.

Valerie Neal recalled, "What I had asked Rutan to do before he delivered it to us was to return it to its June configuration. After that June flight, before the X Prize flights, quite a number of decals were added to it, and the Virgin Galactic logo was added to it. And the appearance was considerably different."

Even the dent in the engine fairing from that flight was put back. That's how seriously Scaled Composites took her request. So, right after *SpaceShipOne* was hung in the museum, the damage drew some quick notice. "The director of the museum came in and said, 'I hope we didn't do that last night.' And I said, 'No, no, it came that way,'" Neal said.

However, even before being transferred to the museum, Rutan wanted to fly *White Knight* and *SpaceShipOne* to Oshkosh. He wanted to do something special for the Experimental Aircraft Association (EAA), which had by him stood from the time of his very first aircraft. With Mike Melvill behind the stick of *White Knight* and *SpaceShipOne* attached below, he flew from Mojave but stopped right before reaching the air show in Madison, Wisconsin, to pick up some very important passengers. Burt Rutan and his wife, Tonya, climbed into *SpaceShipOne*, while Sally Melvill joined her husband in *White Knight*.

They touched down at Oshkosh on June 27, 2005. "It was very emotional because it was like a homecoming for the triumphant soldier," said Tom Poberezny, president of EAA. "And here was Burt coming home to an audience that truly appreciated what he did because they've grown up with him. They've appreciated every design innovation he has ever done, his successes, his failures, his trials, his tribulations."

Figure 10.10 shows the EAA crowd gathered around *SpaceShipOne* and *White Knight*.

When the air show ended, Melvill took off with a small crew to head for Dulles Airport in Washington, D.C., after first stopping over in Dayton, Ohio, the hometown of the Wright brothers. But the adventure was far from over. *White Knight* doesn't have very long legs. Its range is only about 500 miles (800 kilometers). When it reached Dulles Airport, someone must have noticed *White Knight* carrying a missile-like object. "So, they turned us around and drove us away right in the middle of the approach," Mike Melvill said.

"I said, 'If you turn us around, we will run out of gas.' And the air-traffic controller said, 'I don't care. Make a one-eighty and get out of here. I don't want to see you again.' And I said, 'You need to get your supervisor because this has all been pre-briefed.' Pretty soon the airline guys on the same frequency were saying, 'Hey, come on. This is the guy delivering *SpaceShipOne* to the National Air and Space Museum.'"

Even after arrangements had been made with the airport and with the officials from the National Air and Space Museum on the ground, Melvill was denied. But he could always be counted on when the situation did not go exactly as planned. With almost no gas, he was able to land on a runway that wasn't being used by the airlines. After detaching *SpaceShipOne* from *White Knight* and spending about an hour on the ground, Melvill lifted off in *White Knight*. The mothership had left its baby for good. Figure 10.11 shows *SpaceShipOne* in the Milestones of Flight gallery after the donation ceremony on October 5, 2005, hanging next to *Spirit of St. Louis* and *Glamorous Glennis*.

Although *SpaceShipOne*'s mission was suborbital spaceflight, it was actually able to completely break away from Earth's gravitational pull. In 2007, a small piece of *SpaceShipOne*, aboard the space probe *New Horizons*, zipped by Jupiter on its way to a rendezvous with Pluto and its moon Charon. It will then continue on further to the edge of the Solar System into the mysterious Kuiper Belt, a region of space responsible for the demotion of Pluto from a planet to a dwarf planet after the discovery of a tenth planet. Launched in 2006, this is the first mission aimed at exploring these celestial objects. Figure 10.10 shows a conceptual drawing of the space probe on its journey.

In June of 2015, *New Horizons* and the *SpaceShipOne* fragment will have completed the interplanetary cruise phase on the way to Pluto. Earth will be 3.06 billion miles (4.92 billion kilometers) away when the closest approach occurs. Eleven years earlier, to the month, *SpaceShipOne* had first entered space, giving real hope to those with dreams of floating free in space.

A character in Clarke's *2010: Odyssey Two*, in summarizing what he expected from an upcoming space trip, simply stated, "Something wonderful." By the time *New Horizons* actually reaches Pluto, that phrase will be invoked many times thanks to the accomplishments of commercial space travel that are to come.

A: SpaceShipOne Flight Data

Date	Intended Mission	SpaceShipOne										Flight No.[2]	Pilot/ Flight Engineer	Flight Time {hours}
		Flight No.[2]	Pilot	Flight Time {minutes}	Release Altitude {feet (meters)}	Release Speed {knots}	Top Speed {Mach}[b]	Rocket Burn {seconds}	Shutdown Altitude {feet (meters)}	Apogee {feet (meters)}	Maximum g-Force {G}[b]			
5/20/03	Captive Carry	01C										24C	Pete Siebold/ Brian Binnie	1.8
7/29/03	Captive Carry	02C	Mike Melvill									29C	Brian Binnie/ Cory Bird	2.1
8/7/03	Captive Carry	03G	Mike Melvill	19.00	47,000 (14,330)	105						30L	Brian Binnie/ Cory Bird	1.1
8/27/03	Glide[e]	04GC	Mike Melvill									31LC	Brian Binnie/ Cory Bird	1.1
8/27/03	Glide	05G	Mike Melvill	10.50	48,200 (14,690)	105						32L	Brian Binnie/ Cory Bird	1.1
9/23/03	Glide	06G	Mike Melvill	12.25	46,800 (14,270)	115						37L	Pete Siebold/ Matt Stinemetze and Jeff Johnson	1.5
10/17/03	Glide	07G	Mike Melvill	17.82	46,200 (14,080)	115						38L	Pete Siebold/ Cory Bird and David Moore	1.1
11/14/03	Glide	08G	Pete Siebold	19.92	47,300 (14,420)	115						40L	Brian Binnie/ Matt Stinemetze	1.4
11/19/03	Glide	09G	Mike Melvill	12.42	48,300 (14,720)	115						41L	Brian Binnie/ Cory Bird	2.1
12/4/03	Glide	10G	Brian Binnie	13.23	48,400 (14,750)	115						42L	Pete Siebold/ Matt Stinemetze	1.3
12/17/03	Powered	11P	Brian Binnie	18.17	47,900 (14,600)	112	1.2	15	(d)	67,800 (20,670)	3+	43L	Pete Siebold/ Cory Bird	1.2
3/11/04	Glide	12G	Pete Siebold	18.50	48,500 (14,780)	125						49L	Brian Binnie/ Matt Stinemetze	1.3
4/8/04	Powered	13P	Pete Siebold	16.45	45,600 (13,900)	125	1.6	40	(d)	105,000 (32,000)	(d)	53L	Brian Binnie/ Matt Stinemetze	1.3
5/13/04	Powered	14P	Mike Melvill	20.73	46,000 (14,020)	120	2.5	55	150,000 (45,720)	211,400 (64,430)	3.5	56L	Brian Binnie/ Matt Stinemetze	1.5
6/21/04	Powered	15P	Mike Melvill	24.08	47,000 (14,330)	(d)	2.9	76	180,000 (54,860)	328,491 (100,124)	5.0	60L	Brian Binnie/ Matt Stinemetze	1.6
9/29/04	Powered	16P (X1)	Mike Melvill	24.00	46,500 (14,170)	(d)	3.0	77	180,000 (54,860)	337,00 (102,900)	5.1	65L	Brian Binnie/ Matt Stinemetze	1.6
10/4/04	Powered	17P (X2)	Brian Binnie	24.00	47,000 (14,360)	(d)	3.25	84[e]	213,000 (64,920)	367,500 (112,00)	5.4	66L	Mike Melvill/ Matt Stinemetze	1.6

(a) C, G, L, and P denote captive carry, glide, launch, and powered, respectively, for the intended missions of *SpaceShipOne* and *White Knight*. A second letter in the flight number indicates the actual mission if different than the intended mission.
(b) The highest value is given whether occurring during boost or reentry.
(c) Flight aborted prior to *SpaceShipOne* separation from *White Knight*, so *SpaceShipOne* was not released.
(d) Data not reported in Combined *White Knight*/*SpaceShipOne* Flight Tests provided by Scaled Composites.
(e) The value of 84 seconds is used based upon the transcript of 17P.

B: Chase Plane Crews

Flight No.	Duchess: Low Altitude	Extra 300: High Altitude	Alpha Jet: High Altitude	Starship: High Altitude
01C	(a)	(a)	(a)	(a)
02C	(a)	(a)	(a)	(b)
03G	(a)	(a)	(a)	(b)
04GC	Jon Karkow			Pete Siebold
05G	Jon Karkow			Pete Siebold
06G	Brian Binnie			Jon Karkow
07G		Chuck Coleman		Brian Binnie
08G	Mike Melvill Chuck Coleman			Jon Karkow
09G	Chuck Coleman Matt Stinemetze			Pete Siebold
10G		Mike Melvill Chuck Coleman	Marc de van der Shueren Jeff Johnson	Jon Karkow
11P		Mike Melvill Chuck Coleman	Marc de van der Shueren Jeff Johnson	Jon Karkow
12G		Mike Melvill Chuck Coleman		Jon Karkow
13P		Mike Melvill Chuck Coleman	Marc de van der Shueren Jeff Johnson	Jon Karkow Robert Scherer
14P	Pete Siebold Dave Moore		Marc de van der Shueren Jeff Johnson	
15P		Chuck Coleman Cory Bird	Marc de van der Shueren Jeff Johnson	Jon Karkow Robert Scherer
16P		Chuck Coleman Cory Bird	Marc de van der Shueren Jeff Johnson	Jon Karkow Robert Scherer
17P		Chuck Coleman Cory Bird	Marc de van der Shueren Jeff Johnson	Jon Karkow Robert Scherer

(a) Data not reported in *SpaceShipOne/White Knight* Flight Log.
(b) The Starship, owned by Robert Scherer, was flown during this flight, but the crew was not reported.

Selected Bibliography

Books

Belfiore, Michael. *Rocketeers: How a Visionary Band of Business Leaders, Engineers, and Pilots Is Boldly Privatizing Space.* New York: Collins, 2007.

Clary, David A. *Rocket Man: Robert H. Goddard and the Birth of the Space Age.* New York: Hyperion, 2003.

Lindbergh, Charles A. *The Spirit of St. Louis.* New York: Charles Scribner's Sons, 1953.

Neufeld, Michael J. *The Rocket and the Reich: Peenemünde and the Coming of the Ballistic Missile Era.* New York: The Free Press, 1995.

Interviews (with author)

Allen, Paul. Phone interview, October 18, 2007.

Ansari, Anousheh. Phone interview, June 6, 2007.

Bennett, Steve. Phone interview, May 31, 2007.

Binnie, Brian. Phone interview, July 20, 2007.

Diamandis, Peter. Phone interview, November 29, 2006.

Feeney, Brian. Phone interview, May 28, 2007.

Lindbergh, Erik. Phone interview, May 25, 2007.

Maryniak, Gregg. Phone interviews, June 12, 2007, and June 15, 2007.

Melvill, Mike. Interview at Scaled Composites, Mojave, California, June 25, 2007.

Moore, Dave. Interview at Vulcan, Seattle, Washington, July 10, 2007.

Neal, Valerie. Phone interview, June 7, 2007.

Poberezny, Tom. Phone interview, June 26, 2007.

Rutan, Burt. Interviews at Scaled Composites, Mojave, California, June 25, 2007, and June 26, 2007.

Shane, Doug. Interview at Scaled Composites, Mojave, California, June 26, 2007.

Smith, Patty Grace. Phone interview, June 11, 2007.

Siebold, Pete. Interview at Scaled Composites, Mojave, California, June 25, 2007.

Tucker, Sean D. Phone interview, June 1, 2007.

Magazine articles

Binnie, Brian. "Confessions of a Spaceship Pilot." *Air & Space*, June/July 2005.

Dornheim, Michael A. "Affordable Spaceship." *Aviation Week & Space Technology*, April 21, 2003.

Miller, Kenneth. "Your Spaceship Awaits." *Life*, October 22, 2004.

Parker, Ian. "The X Prize." *The New Yorker*, October 4, 2004.

Sweetman, Bill. "*SpaceShipOne*: Riding a *White Knight* to Space." *Aerospace America*, American Institute of Aeronautics and Astronautics, January 2004.

Thomas, Cathy Booth. "The Space Cowboys." *Time*, March 5, 2007.

Documents and Papers

NASA. "Dryden Flight Research Center." Fact sheet FS-2003-05-001-DFRC, May 2003.

NASA. "Major NASA Launches." Information summary PMS 031 (KSC), June 1999.

NASA. "NASA's Orbiter Fleet." Fact sheet FS-2007-07-025-KSC, July 2007.

NASA. "New Horizons, The First Mission to Pluto and the Kuiper Belt: Exploring Frontier Worlds." Launch press kit, January 2006.

NASA. "Space Shuttle Launches." Information summary IS-2004-03-003-KSC, March 2004.

NASA. "Space Shuttle Mission Chronology, Vol. 1, 1981–1999." Information summary NP-1997-12-08-KSC-REV, January 2000.

NASA. "Space Shuttle Mission Chronology, Vol. 2, 2000–2003." Information summary NP-2005-03-02-KSC, March 2005.

NASA. "Space Shuttle Propulsion Systems." Fact sheet FS-2005-04-028-MSFC, April 2005.

NASA. "Space Shuttle Use of Propellants and Fluids." Fact sheet FS-2001-09-015-KSC, September 2001.

X Prize Foundation. "Ansari X Prize Competition Launches." Commemorative program, 2004.

X Prize Foundation. "Ansari X Prize Team Summary Sheet: Mojave Aerospace Ventures Team." Media information, 2004.

X Prize Foundation. "Team Descriptions: Ansari X Prize Pioneers of Commercial Space Travel." Media information, 2004.

Scaled Composites. "Flight Navigation Unit." Fact sheet, 2003.

Scaled Composites. "Mission Control." Fact sheet, 2003.

Scaled Composites. "Oxidizer Tank and CTN." Fact sheet, 2003.

Scaled Composites. "Propulsion Test Trailer." Fact sheet, 2003.

Scaled Composites. "Simulator." Fact sheet, 2003.

Scaled Composites. "*SpaceShipOne* Flight." Fact sheet, 2003.

Scaled Composites. "*SpaceShipOne*: The First Non-Government Manned Spacecraft." Fact sheet, 2003.

Scaled Composites. "*SpaceShipOne* . . . The First Private Manned Space Program Goal Is Affordable Sub-orbital Space Flight." Fact sheet, 2003.

Scaled Composites. "*White Knight*." Fact sheet, 2003.

Scaled Composites. "*White Knight* . . . A New High-Altitude Research Aircraft by Scaled Composites." Fact sheet, 2003.

Virgin Galactic. "Virgin Galactic Showcases its Investment in the Development of SpaceShipTwo." Media information, September 28, 2006.

Web articles

Boyle, Alan. "Spaceship Team Gets Its $10 Million Prize." *MSNBC* (November 6, 2004), http://www.msnbc.msn.com/id/6421889 (accessed September 5, 2007).

Dash, Eric, et al. "America's 40 Richest Under 40." *Fortune* (September 17, 2001), http://money.cnn.com/magazines/fortune/fortune_archive/2001/09/17/310275/index.htm (accessed June 5, 2007).

David, Grainger. "Boldest Newcomer." *Fortune* (September 17, 2001), http://money.cnn.com/magazines/fortune/fortune_archive/2001/09/17/310261/index.htm (accessed May 11, 2007).

David, Leonard. "*SpaceShipOne* Makes History with First Manned Private Spaceflight." *Space.com* (June 21, 2004), http://www.space.com/missionlaunches/SS1_press_040621.html (accessed May 5, 2007).

David, Leonard. "*SpaceShipOne* Wins $10 Million Ansari X Prize in Historic 2nd Trip to Space." *Space.com* (October 4, 2004), http://www.space.com/missionlaunches/xprize2_success_041004.html (accessed October 2, 2007).

David, Leonard. "The Next Great Space Race: *SpaceShipOne* and *Wild Fire* to Go for the Gold." *Space.com* (July 27, 2004), http://www.space.com/missionlaunches/xprize_spacerace_archive.html (accessed May 23, 2007).

de Córdoba, S. Sanz Fernández. "100 km. Altitude Boundary for Astronautics . . . " *Fédération Aéronautique Internationale* (June 21, 2004), http://www.fai.org/astronautics/100km.asp (accessed June 17, 2007).

Foust, Jeff. "Burt Rutan, in His Own Words." *The Space Review* (October 25, 2004), http://www.thespacereview.com/article/255/1 and http://www.thespacereview.com/article/255/2 (accessed May 9, 2007).

Klotz, Irene Mona. "Virgin Soars Towards New Frontier." *BBC News* (December 27, 2004), http://news.bbc.co.uk/2/hi/science/nature/4119491.stm (accessed September 6, 2007).

Miller, Megan. "A Giant Leap—and Crash—for the Lunar Lander Challenge." *PopSci.com* (October 2006), http://www.popsci.com/popsci/aviationspace/6fbcb6d013c6e010vgnvcm1000004eecbccdrcrd.html (accessed September 21, 2007).

Snelson, Robin. "Unsung Heroes of the Personal Spaceflight Revolution." *The Space Review* (September 27, 2004), http://www.thespacereview.com/article/234/1 (accessed September 6, 2007).

Valdes, Robert. "How *SpaceShipOne* Works." *HowStuffWorks*, http://science.howstuffworks.com/spaceshipone.htm (accessed October 8, 2006).

Websites

CharlesLindbergh.com. "The Flight." http://www.charleslindbergh.com/history/paris.asp (accessed May 22, 2007).

da Vinci Project. "Worlds Largest Reusable Helium Balloon Completed by Canadian Manned Space Flight Team." December 12, 2004, http://www.davinciproject.com/beta/News/NewsMain.html (accessed September 3, 2007).

EAA. "Lindbergh Grandson to Follow Grandfather's Flightpath." http://www.eaa.org/communications/eaanews/020208_lindbergh.html (accessed May 27, 2007).

EAA AirVenture. "Hallmarks of Homebuilding: Ken Rand's Composite Airplane, The KR-1." http://www.airventure.org/2006/events/hallmarks_rand5.html (accessed May 17, 2007).

EAA AirVenture Museum. "EAA/Scaled Composites *SpaceShipOne*—Replica." http://www.airventuremuseum.org/collection/aircraft/SpaceShipOne.asp (accessed September 21, 2007).

EAA AirVenture Museum. "Scaled Composites/Rutan Voyager Fuselage Replica." http://www.airventuremuseum.org/collection/aircraft/Scaled%20Composites-Rutan%20Voyager.asp (accessed May 20, 2007).

NASA. "A History of Human Spaceflight." http://www.nasa.gov/multimedia/imagegallery/image_feature_800.html (accessed October 2, 2007).

NASA. "About the International Space Station." http://pdlprod3.hosc.msfc.nasa.gov/D-aboutiss/index.html (accessed September 3, 2007).

NASA. "Speed of Sound." http://www.grc.nasa.gov/WWW/K-12/airplane/sound.html (accessed November 3, 2007).

NASA. "Sputnik 1." http://nssdc.gsfc.nasa.gov/database/MasterCatalog?sc=1957-001B (accessed September 3, 2007).

NASA. "X-15." http://www.nasa.gov/centers/dryden/about/Organizations/Technology/Facts/TF-2004-16-DFRC_prt.htm (accessed April 27, 2007).

NASA. "X-15 Launch from B-52 Mothership." http://www1.dfrc.nasa.gov/Gallery/Photo/X-15/HTML/E-4942.html (accessed June 21, 2007).

National Air & Space Museum. "Bell X-1 *Glamorous Glennis*." http://www.nasm.si.edu/exhibitions/gal100/bellX1.html (accessed November 25, 2006).

National Air & Space Museum. "*SpaceShipOne*." http://www.nasm.si.edu/exhibitions/gal100/ss1.htm (accessed October 12, 2006).

National Air & Space Museum. "*SpaceShipOne* Joins the Icons of Flight on Display at Smithsonian's National Air and Space Museum." (October 5, 2005), http://www.nasm.si.edu/events/pressroom/releaseDetail.cfm?releaseID=138 (accessed October 2, 2007).

Scaled Composites. "Combined *White Knight*/*SpaceShipOne* Flight Tests." http://www.scaled.com/projects/tierone/logs-WK-SS1.htm (accessed September 26, 2006).

Scaled Composites. "Frequently Asked Questions—General." http://www.scaled.com/projects/tierone/faq.htm (accessed April 21, 2007).

Scaled Composites. "Our Message at the April 18th 2003 Unveiling of the Tier One Program." http://www.scaled.com/projects/tierone/message.htm (accessed September 26, 2006).

Scaled Composites. "Paul G. Allen Confirmed as Long-Rumored Sponsor of *SpaceShipOne*." (December 17, 2003), http://www.scaled.com/projects/tierone/121803.htm (accessed October 13, 2006)

Scaled Composites. "Proteus, A High-Altitude, Multi-Mission Aircraft." http://www.scaled.com/projects/proteus.html (accessed September 9, 2007).

Scaled Composites. "Scaled Composites Unveils the Existence of a Commercial Manned Space Program." (April 18, 2003), http://www.scaled.com/projects/tierone/041803.htm (accessed October 10, 2006).

Scaled Composites. "*SpaceShipOne* Breaks the Sound Barrier." (December 17, 2003), http://www.scaled.com/projects/tierone/121703.htm (accessed May 4, 2007).

Scaled Composites. "*SpaceShipOne* Makes History: First Private Manned Mission to Space." (June 21, 2004), http://www.scaled.com/projects/tierone/062104-2.htm (accessed May 5, 2007).

Scaled Composites. "*White Knight* Flight Test Summaries." http://www.scaled.com/projects/tierone/logs-WK.htm (accessed October 13, 2006).

Starchaser Industries. "Chronology of Key Events." http://www.starchaser.co.uk/index.php?view=chronology (accessed May 29, 2007).

Starchaser Industries. "*Nova*/Starchaser 4." http://www.starchaser.co.uk/index.php?view=starchaser4_project&mgroup=projects (accessed September 3, 2007).

U.S. Centennial of Flight Commission. "Early Reentry Vehicles: Blunt Bodies and Ablatives." http://www.centennialofflight.gov/essay/Evolution_of_Technology/reentry/Tech19.htm (accessed April 28, 2007).

X Prize Foundation. "Rules and Guidelines." http://web1-xprize.primary.net/teams/rules_and_guidelines.php (accessed November 25, 2006).

Other

Discovery Channel. *Black Sky: The Race for Space and Winning the X-Prize*. Video documentary, 2005.

Mojave Aerospace Ventures. "Winged Spacecraft." United States Patent, patent number 7,195,207; Date of patent: March 27, 2007.

Shane, Doug. "Risk Management in the Deep End of the Pool—Winning the X-Prize." Presented at the director's colloquium, NASA Ames Research Center, Moffett Field, California, June 6, 2007.

Index